Static! Designing for energy awareness

Ramia Mazé

Static! Designing for energy awareness

Edited by
Ramia Mazé

Commissioned by
The Interactive Institute
The Nordic Culture Fund
The Swedish Energy Agency

Photographs by
Carl Dahlstedt pp. 54, 57-59, 63-64, 67, 74, 77, 82, 84, 86-89;
Remaining photographs and images by the project team.

Design by
Christian Altmann
Oskar Holmquist

Printed by
Printografen AB
Helsingborg 2010

Paper
Cover: Invercote 300g
Text: Munken print white 15 115g

Font
Akkurat Bold/Light/Light italics

Published by
Arvinius Förlag
Box 6040
102 31 Stockholm

www.arvinius.se

ISBN 978-91-85689-34-7

INTERACTIVE INSTITUTE

Nordic Culture Fund

Contents

Static! Designing for energy awareness

Much of our everyday lives is powered by electricity. From the boost of the first cup of coffee in the morning to the soft glow of a nightlight that keeps a child company after dark, from the devices that keep us informed and entertained to the treasured decorations that illuminate special occasions... these things, and the activities and rituals that accompany our use of them, demonstrate that electricity consumption is intimately interwoven into personal and domestic life. Today, however, the economic and environmental cost of energy forces us to think again.

Electricity is often perceived as abstract – it is hard to grasp the amount or consequences of electricity consumed at any given time or through any particular product. Did you know that up to 20% of household electricity bills is for producing hot water? 10% for lighting? That products may continue to draw up to 85% of their normal power consumption when in standby mode? In domestic life, consumption is often only present once a month, in the complex and quantitative form of the monthly electricity bill, perhaps viewed only by the head of the household. If the costs and consequences of electricity – not to mention our habits and trends in consumption – remain disconnected from everyday life, electricity may only too easily continue to be perceived as unlimited and always on tap.

In rethinking energy consumption, we must also rethink the things through which we consume energy. But the design of everyday things does not often express much about energy. The infrastructure for delivering electricity and the meters for measuring it are often invisible, hidden away behind wallpapered surfaces or neat packaging. Indeed, aesthetics have long been applied to render product – and electricity – consumption effortless, pleasurable and desirable. In fact, design has been complicit in increasing consumption – consider, for example, the introduction of electrical appliances for domestic use in the early 20th century. Appliances such as electric kettles, irons and washing machines by Peter Behrens, often considered to be the first industrial designer, succeeded in dramatically increasing general electricity consumption on behalf his client AEG, a German electricity company.

The disciplines of industrial design and, more recently, interaction design have grown up alongside interest in increasing the profitability of emerging electric and electronics sectors – but, today, the challenge for design is to change behavioral patterns of energy (over)consumption.

Sustainability is rapidly transforming the culture of design. But, even if it may appear so when printed in our monthly bill, electricity consumption is not a black and white – or only a green – issue. Consider that the first electrical appliances, for example, also accompanied a societal revolution, changing the nature and perception of 'domestic labor' and shifting established class and gender roles. Hot water and central heating systems help ensure hygiene and comfort, bound up into quality of life and family welfare issues. Electricity consumption, thus, is not only a question of economic cost to be quantified and reduced – but a question of social and cultural values. Relations between design and energy are complex, each bound up with long histories and meaningful traditions – and each a powerful force in changing our future.

Today, we need to forge new relations between design and energy, and not only on an industrial or global scale. Indeed, the debate on sustainability is tied to large ethical and ideological issues in society that are already too hard to relate to our individual actions and local choices. Rooted in the design of everyday life, design might redirect its aesthetics and strategies towards new ends. If there cannot be one solution, or even an end game, to energy futures, then we might start here and now, with what we have and what we know. For designers, this means critically rethinking processes and products to forge a more responsible relationship to clients and consumers. For energy consumers – that is, all of us – this means reconnecting to the societal and humanitarian implications of energy consumption by starting with the electricity in the things that we already use and value in everyday life.

While rhetorics around energy (over)consumption are all too easily reduced to the simplistic terms of 'on/off', this book explores shades of grey in between. Instead of focusing solely on preserving nature or conserving energy, per se, the politics of sustainability are confronted with the poetics of design and everyday life. Through the lenses of critical and user-centered design, we question mainstream tendencies of production and consumption, arguing for humanistic values and societal leadership by design. Through one of our design research projects, we illustrate how the challenges of sustainability – and, in particular, issues of energy consumption – have reconfigured how we think about design practice and design research, about the consumers and consequences of design.

'Static!' is a design research project, led by the Interactive Institute and funded by the Swedish Energy Agency, that investigated how the design of things we use everyday might be a basis for increasing awareness of energy. In the project, we sought a middle ground between the complex societal debate and the simplistic terms – on/off, consume/conserve, save/spend – to which such debate is often reduced. As an alternative, we investigated the power of product design to materialize energy – to render it more visible and experiential in everyday life – and thus increase reflection and choice in daily life and lifestyles. Design research has been our platform for communicating with users and designers – for supporting reflection in use and for developing a more profound understanding of energy in design.

A collection of design examples was produced in Static!, including prototypes, conceptual design proposals, and use scenarios. These exemplify alternatives ways of materializing and interacting with energy through everyday things. Aesthetics are developed to render energy more visible and tangible – more available to sensory experience. Thus, energy is not something merely to optimize or hide away but an essential and expressive material. Besides developing alternative aesthetics of energy, we designed interactions with energy through products over time, such that repeated and ongoing use might affect energy behaviors. Our research approach builds on two main ideas: that we, as designers, can work with energy not only from a technical but also from an aesthetic point of view, thereby integrating the often separate areas of design and engineering and that product use need not only be about utility and ease-of-use but also about critical reflection on energy through the objects at hand.

Static! is anchored in different disciplines – including service and industrial design, furniture and textile design, industrial design and engineering, conceptual and critical design. Some design examples focus on visualization and efficiency, others materialize questions about the status quo, relating to values of emotion, memory, ethics, and persuasion. Still others employ design as a vehicle for carrying out a participatory discussion about possible futures with various groups. Diverse approaches stake out a rich spectrum of ideas, values, and voices present in current debates about energy and within design. Deployment of design examples spans from long-term evaluations in households and surveys of commercial prospects, to debates on design and energy in local and expert forums. Outcomes in the form of publications and exhibitions have been targeted to impact various stakeholders, including industry, municipal agencies, academia, media, and the public sector.

In this book, we describe some motivations, methods, and outcomes of Static! First, there is an introduction to the strategic and research setup of the project, closely followed by an essay that relates the research themes to bigger issues in design and sustainability discourse. Perspectives on our relation to certain tendencies in contemporary design practice include people-centered and critical design. Then, the design examples are explained and illustrated at length. Reflecting on the reception of the examples in household studies and of our research upon various stakeholders, to culminate, we discuss the impact and new directions of the work.

Collectively, the perspectives and examples in this book map an emerging design space – a territory of possible relations to energy in design that we have only begun to investigate, challenge, and change. Grounded in passion and humor, as well as rigor and research, this situates designers and users of energy – that is, all of us – as engaged participants in changing the form, and future, of the world around us.

Strategic frames

Christian Öhman and
Sara Backlund

The Interactive Institute is a non-profit research institute, in which technology is investigated and developed from the perspectives of art and design. Founded in 1998 by the Swedish Foundation for Strategic Research, the institute consists of studios distributed within Sweden, each with its own research theme and associated partners. Operating between industrial, academic, cultural and public sectors, the ambition is to create a more inclusive, creative and innovative arena for research. Indeed, the institute is explicitly set up to encourage experimental initiatives that cannot be readily undertaken within a company or university alone. Through research programs, public productions, entrepreneurial activities and networks, we work collaboratively to develop new ways of thinking about and applying technology.

Research at the Interactive Institute is often associated with making the 'invisible' more apparent. Building physical prototypes and creating concepts that exemplify abstract theories or complex systems are important aspects of our work. This has proven to be especially useful when it comes to understanding the relation between design and energy. In Static!, the theme 'energy as a design material' distilled both a research issue and design challenge into a strategy for making something that is ordinarily rather abstract into something that is visual and tangible – to designers, consumers, and other stakeholders. In demonstrating a range of ways to materialize the presence and use of energy in everyday domestic life, resulting design examples have, in turn, produced a range of positive results and success stories.

The creation of the Static! project in 2004 was the first research initiative to develop out of a growing relationship between the institute and the Swedish Energy Agency. The agency had approximately 40 research programs in areas such as systems analysis, building infrastructures, transportation, and energy-intensive industries. There had been only a few small initiatives focused on consumers – indeed, it is only much more recently, for instance, as the climate has entered public consciousness in a big way, that the 'soft power' of such approaches to change perception and behavior has become apparent. Ahead of this trend, the agency was extremely foresighted to venture outside its established research areas – into a relationship with the Interactive Institute and a conversation in which human needs and creative practice would take precedence over technology. In particular, Andres Muld took a particular interest in the startup of a new studio within the institute focused on – and named – POWER.

Because both organizations were branching into an area in which expectations and relations had yet to be established, it was an important step to formulate and carry out a research project. This project became a platform for facilitating a mutual learning experience and growing our agendas together. Under the management of Christina Öhman, who initiated the POWER studio, the Static! project and research scope was formulated under the research direction of Johan Redström and Ramia Mazé from another more established studio in the institute focused on design. Alongside gathering various competences within the institute into a multidisciplinary team of about 15 people, we also invited a number of postgraduate and doctoral students to participate and, significantly, the newly started Front Design group. The team was the foundation for a content- and practice-based inquiry.

There were several organizational factors built into Static! in order to facilitate evolution of a new research area and its unanticipated potentials during the project period. The work was conducted as a series of design loops, in which design examples were developed within smaller teams, operating independently and in parallel with one another, with periodic phases of convergence. Early in the project, convergence involved internal activities for us to communicate and synthesize the ideas, as a basis for continuing or redirecting work during the next loop. At key points, we invited experts from universities and from the Energy Agency to evaluate the work from specific research perspectives. Later, as the design examples became more resolved, we identified further competences needed and developed relationships with additional partners in academia, manufacturing, and enterprise. This open-ended structure allowed for both independent and collective interests, as well as for the introduction of diverse inputs and stakeholders along the way.

While the individual design examples created within Static! may be the most visible results, these are not intended as ends in themselves. Behind our efforts at the Interactive Institute to make visible abstract or complex technologies is an ambition to make these technologies more accessible to a wider audience. Through examples of potential applications and scenarios of potential use, the implications of technical systems become more understandable and future lifestyles more imaginable. Unlike prototypes in professional design practice, the design examples in Static! are not intended as final solutions or even product proposals. Static! is less about design for mass-production or even design for actual use – instead, 'design for debate' better characterizes our use of design to reach out to new audiences.

Designing for energy awareness

Rather than means or ends in and of themselves, the processes and 'products' of Static! were intended as open-ends – to invite further discussion and participation with potential stakeholders. In this case, the audience – or 'users' – of design research included not only potential end-users, but also a range of stakeholders in academia and industry, various creative and scientific disciplines, as well as institutions in the public sector including our sponsor. The design examples, taken collectively, opened up for staging a discussion around alternative approaches to energy awareness with a wider audience. Beyond and after the creation of the design examples, this required further attention to the creation of formats of outreach initiatives as well as forms for articulating and disseminating resulting knowledge.

Indeed, experimenting with new formats for involving stakeholders has been essential for evaluating, developing and disseminating the work. One example is our 'Energy+Design network', funded by the municipality. Discovering a range of complementary – and potentially competing – interests in sustainability, we created a network to open a more participatory discussion. In a series of seminars, and with Static! as a backdrop, ethical, commercial and societal interests were exposed through invited speeches and public debates. The Static! 'Energy Curtain' is another example of stakeholder involvement. While first developed as a conceptual design, its potential feasibility was developed together with the textile manufacturer Ludvig Svensson, the microelectronics institute Imego, and a startup lighting company. These initiatives demonstrate an intersection of research activities and communications formats, orchestrated within and around the project, to spark cross-sector collaboration.

We are also actively communicating with the varied audiences and beneficiaries of such design research. Partners have benefited from participating in the development of the design examples – in terms of potential commercial outcomes as well as knowledge about design research and collaboration methods. With respect to consumption, two extended studies of three design examples in multiple households have provided insight into user perceptions and behavioral changes. Exhibition in diverse contexts, from art and design museums to energy and technology expos, have opened up for different forms of influence, dialog and criticism. In addition to extensive exposure in the media, authored publications in international journals and conferences have expanded upon the theoretical and methodological results. From networks and collaborations, user studies and public arenas, peer-review forums and mass media, outreach – and impact – has taken many forms. 'Testing' ideas and outcomes in relation to diverse audiences has staked out a vital arena for carrying out and reflecting upon Static!, as well as for launching future work.

Research frames
Johan Redström

It might, at times, seem as if design theory and design practice exist quite far from one another. The former, in particular, might seem to lead a rather independent life, distant from the pragmatics of daily operation. Closer inspection, however, often reveals the opposite – that the theories and questions occupying research are quite intimately tied to the context in which they are developed. Indeed, the approaches and developments in Static! were responsive to various disciplinary, institutional and national conditions. For this pilot project with the Swedish Energy Agency, for example, the ambition was to do something exemplary, even pedagogical, to introduce design research and produce a variety of representative results. Further, it was meant to bootstrap the development of a new research studio by, in part, building on research established elsewhere at the Interactive Institute. An initial challenge, thus, was to form a coherent theoretical and operational program for the research.

Static! was originally configured along two parallel tracks, each hosted by a research studio. One would build on established research in experimental interaction design, investigating energy in terms of materiality and aesthetics. The other would investigate the influence of design concepts and prototypes on people and their energy behaviors. This can be seen in early texts such as the abstract of the original funding application:

Static! investigates interaction design as a way of increasing our awareness of how energy is used and how to stimulate changes in energy behavior. Revisiting the design of everyday things with focus on issues related to energy use, we will develop a palette of critical design examples, e.g., prototypes, conceptual design proposals and use scenarios. These design examples will then be used as a basis for communication and discussion with users and designers, for developing a more profound understanding of energy in design, and to support awareness of design issues related to energy use early in the product development process.

Already in the first sentence, the two perspectives are evident. Indeed, there is an ambivalence between an open investigation of design as a means of exposing issues related to energy, and a more instrumental understanding of design as a way to investigate and influence energy-related behaviors. In practice, the two tracks did not really come about, and it is probably fair to say the program failed to develop the depth aimed for in terms of studying user behavior. On the other hand, the palette of design examples was much more rich, diverse and successful than originally hoped, since both groups were engaged in their development. Indeed, perspectives ranging across contemporary design were present – from the applied arts to human-computer interaction, from media art and activism to service and strategic design.

The problematics of framing research in Static! is perhaps characteristic of the situation for design research in general. It can be easy to charge design research with taking on far too large topics, spanning far too many issues along the way and, thus, risking lack of precision in the end. A typical response to such a charge is that designers and design researchers are 'generalists', 'integrators', or 'divergent thinkers' rather than the kind of experts more typical in traditional research. There are also other reasons behind such broad agendas. Since there is little research funding targeted explicitly towards design, design research must operate in relation to the different, and often competing, terms of industry and academia. At a further intersection within academia – between the arts and humanities, the social and technical sciences – design research often has to shift between, or bridge across, disciplines and domains.

To deal with these conditions – both on the ground and inherent in the field – we constructed a research program for Static! By research 'program', we refer to a set of theoretical concepts and working methods that act as a common ground for enquiry and experimentation combining diverse perspectives, disciplines, and stakeholders. Within a tentative or provisional knowledge regime framed by a program, the role of theory is to locate a discursive context for experimentation with materials, technologies, scenarios, methods and prototypes. Importantly, this does not imply the resolution of difference within one grand unified theory – indeed, the task of interpreting and negotiating diverse matters of concern is, to a significant extent, situated within the localized practices involved in creating the design examples.

While a research program therefore deals with a range of issues – some of which might appear only remotely related to 'real' theory or practice – negotiating such issues exposes a range of new problems and possibilities, since (design) research is not only about poetics – but often also about politics.

Research program
The research program in Static! builds on a range of perspectives, including such diverse approaches as critical design, persuasive design and sustainable design as well as the aesthetics of materials, technology and interaction. The basic idea of 'design for energy awareness' is one expression combining two different ambitions – one directed towards design and designers, the other towards use and users.

On one hand, the program is directed inwards to design practice, exploring how designers can engage with energy as a material rather than as an abstract or invisible technology. Essential to building the appearance and behaviors of products and environments, the presence of energy in everyday life can become something to visualize and materialize more explicitly. On the other hand, the program is directed outwards towards users, deploying design to express issues related to energy use through the form of and interaction with everyday things. Materializing energy and energy use, design examples can become vehicles for stimulating reflection among users.

These ambitions can be further articulated as an investigation into:

- The aesthetics of energy as material in design – working with energy not only from a technical but also from an aesthetic point of view
- Reflective use – systematically reinterpreting designed things not only in terms of utility and ease-of-use but in terms of critical reflection through the things at hand

The first proposition articulates a relation to technology in Static! Here, technology is not understood simply as the means to an end, an end typically referring to a solution to (all) our problems. Instead, technology is approached as a material with certain characteristic and integral expressions.

The second proposition articulates a position with respect to utility. While 'use' can be influenced and potentially transformed through design, criticality and curiosity might be strategies for designing everyday things to incite reflection upon the presence and consumption of energy.

Our hope, on the basis of these ambitions, was that expressing energy as integral to the use of objects might close the gap between information about energy use and actual behaviors in use. By articulating this in terms of two propositions, we try to express that negotiating this gap depends as much upon developing the knowledge and practices of design research as of potential end-use. Looking 'inwards' and 'outwards' simultaneously, the program sets up a tension for thinking and working between these two interdependent sets of concerns and stakeholders in design research.

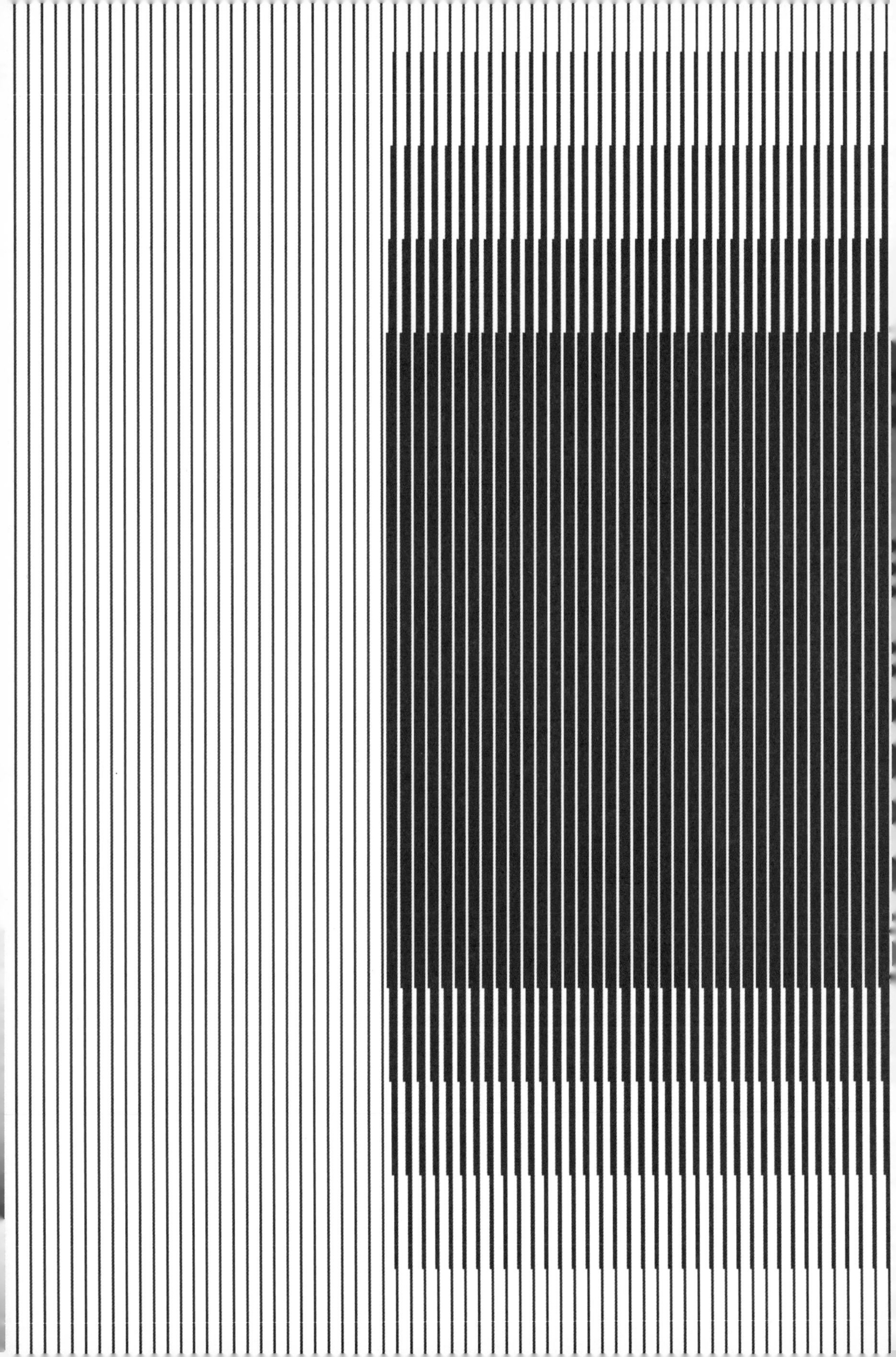

Essay

'Current' issues in sustainable design and research

Johan Redström

Though the heated debate regarding our current (Western) lifestyle and its consequences for the future of the global environment is far from settled, it is increasingly clear that present trends cannot continue unchallenged. The question of sustainability, however, relates to a complex set of issues that are layered and interwoven in ways that few, if any, of us can grasp fully. Even so, our desire for simple solutions to complex problems is as strong as our faith in technological development to resolve them for us.

While design has, for some time, been at the forefront of expanding consumption and material welfare, it now faces the issue of how to achieve, if not the opposite, then, at least, a more conscious and reflective mode of consumption. Obviously, we need to challenge the design tenet of promoting increased consumption through irresistible objects, ownership and identity as well as the commercial strategy of producing objects with short life spans in order to increase consumption over time. But proposals such as ceasing to consume altogether are not viable, since there are not only the ecological aspects of consumption to consider. Indeed, there are rich and often complicated emotional, cultural and existential values embedded in various everyday practices that will not disappear even if hid beneath simplified notions of consumption.[1] The details of this picture cannot be seen if processed in only black or white – we need full color.

At the same time, there is no future in the idea of eternal solutions to temporal problems. The ideal of a kind of timeless design with eternal values that do not go out of fashion is not viable either – not least because of the normative and conservative mechanisms typically associated with such perspectives. Although the future is as uncertain as usual, it seems increasingly clear that there is simply no turning back or reversal of current development trends. Thus, the problem is not about how to make a full stop (or rewind), but about how to initiate change.

Increasing awareness of the dystopic nature of some future scenarios has been followed by an increasing amount of research on how to predict and, perhaps, eventually prevent such scenarios from becoming reality. For design, this not only requires that we revisit basic issues and assumptions, it opens up for creating new visions about the future. The current situation suggests that just maintaining the status quo is no longer an option, that new and 'forceful propositions' about alternative futures is a possibility – and a necessity.[2] While we should not underestimate the gravity of the issues at hand, we should nonetheless appreciate the consequent possibilities for design research and practice. Although these issues may seem to invite it, this is not a time for taking on extreme positions and polarized debates about lost or found utopias, but for acknowledging that there are both problems and potentials in the current situation that require us to engage in diversity and everyday pragmatics. Seriousness and rigor are important, but so are passion and humor.

Here, I outline an approach to this challenge. Though dealing with some more general issues in sustainable design, the approach grew out of our work investigating how interaction design can be used to promote energy awareness in everyday life. The idea of using design to promote energy awareness may sound straightforward enough but, in fact, it is far from obvious how to unpack it. We wanted to avoid the position that it is consumption, somehow considered as a single or measurable thing (rather than as a range of complex and creative practices), which is the problem – a problem that would presumably vanish once people are sufficiently educated. We also wanted to avoid the position that new technology will be the solution, that environmental issues will be resolved once we develop and apply the appropriate technology. We wanted to explore an alternative to the currently dominant approaches of consumer information campaigns and of energy-efficient technology. This is not to say that these approaches to initiating change are not valuable – rather, that we wanted to position ourselves in a way that enabled us to work with energy awareness and that related critically to established positions within design and technology development.

We wanted to create another design space. Although we wanted to relate to consumers and users, technology developers and designers, we approached this in another way – by revisiting and challenging basic conceptions in design. From this perspective, our approach defies the notion of design as a matter of problem solving. If it is 'change' rather than 'solutions' at stake, sustainable design might instead start with questions of engagement and empowerment, rather than with problems. Further, the challenges of sustainable design require us to move beyond design preoccupied with surfaces. Instead of superficially styling or packaging otherwise complex messages or technologies, perhaps we might consider the designed object as an opportunity for unpacking otherwise invisible and hidden processes.

Problem

awareness

Although many agree that change is needed, it less clear precisely what should be done and how it should happen. One seemingly popular approach to sustainable design is the turn towards more ecologically-sound materials and means of production. Such strategies, however, do not necessarily address issues related to lifestyles and (over-)consumption. A good illustration of this problem is the debate around alternative fuels, which is not only about the design and manufacture of hybrid cars but about how local authorities encourage their use through 'eco-car' incentives such as reduced parking fees and road taxes.

Yet another approach to sustainability is to shift from material products to immaterial services, thereby, seemingly, avoiding the problems of producing and consuming material resources. But this can risk ignoring the importance of material life and culture. In the technology domain, visions of the 'paperless office' have argued in a similar way that information and communication technologies would replace material artifacts and even physical meetings. Even if this is taken in light of current sustainability issues, it is not only the material hardware of such technologies, but electrical consumption, that is a significant cost of intensive computing and server farms. Even an e-mail account that I never use leaves an ecological footprint.

Due to the complexity involved in sustainability, it is all too easy to spot the shortcomings of design. It is simply very hard to consider and negotiate all aspects necessary to achieve a fail-safe solution – especially as these solutions are typically meant to be compatible with the current state of affairs. The important thing, however, is to realize that it is not simply a matter of finding better solutions, but that we are facing a problem that it is not up to design to solve – it is unlikely that new kinds of design or technology development would find a way to solve the current problems, thereby allowing us to continue as is. This represents a challenge to the very foundations of design, to our understanding of the objectives and objects of design as solving problems. And, strange as it may sound, it might not be solutions that we should be looking for.

Observing that there will always be more or less significant shortcomings of approaches to sustainable design tells us several important things. First of all, there is no such thing as simple solutions, only more or less viable proposals of what the future could be like – therefore, we must remember to relate to them as such. Secondly, the idea that design and technology will deliver sustainable solutions is a high-risk project, to say the least. When – not if – shortcomings become exposed, we risk a backlash against sustainable development, accompanied by the pessimism and hopelessness of those people who do adopt the new designs and technologies proposed, but to less effect than desired or promised. Further, there is a risk in the suggestion that there will eventually be solutions that allow us to continue as is – that it will be up to new technology to resolve the matter for us. In all of these cases, we risk diffusing engagement, passion and concern – and, thus, the necessary basis for initiating and carrying out significant change. And we are going to need that engagement, because the question of sustainable development is at least as much about lifestyle as it is about design or technology.

So, let's put the idea of problem-solving aside, at least for a while, in order to explore some alternative perspectives. We do not put it to the side because it is not important. However, we also need to revisit basic conceptions in design – it will not be enough to rethink materials used while ignoring the consequences of current thinking about form. It will not be enough to rethink form while ignoring issues of materiality. And, to start, we need to consider how use unfolds over time, and how design might invite and encourage engagement and empowerment.

unpacking

As mentioned, our perspective on time in consumption must shift from a focus on the experience of acquisition to the experience of use and what happens during the life of products. Some such views are already underway, evident, for instance, in efforts to position products as expressions of a certain lifestyle and in strategies to form long-term relationships between a brand and consumer through repeat purchases, services and maintenance programs.[3] But sustainability requires us to shift even further from the emphasis on 'buying a product' towards 'buying into a process'.

It is obvious that a purchased product is but the last step in a long process spanning from the materials made and used to the conditions for production and transportation. However, conceived as an object, rather than a process, information about origin, development and conveyance can easily become seen as merely supplemental, something that might – or, in some cases, might not – add to the perception or value of a product. The use of labeling standards for ecological production and fair-trade can be seen as a response to the fact that such information is typically seen as something external to the object, something that has to be added on because it is not inherently present. In general, some types of information (like brand name and country of production) readily filters through, whereas others (like workers' conditions and environmental impact) typically have more difficulty reaching an end-customer as central to what the product is. Indeed, even the notion of an 'end-customer' says something about the problem, since few products actually end their life in the possession of the one who originally purchased them. Gifts and waste, recycling and second-hand markets, illustrate how far from terminal the end of 'end-customer' actually is.[4]

Further, current practice does not only hide processes involved in the object's past, it also frequently hides aspects of its use – or *conceals*, as some philosophers of technology put it.[5] The assumption is that design should, ideally, not only make it easy to fall in love with things, it should also make what follows afterwards as convenient as possible. Particularly in the design of technology, there have been substantial efforts to make complex products easy to understand and use, often by presenting a user with a relatively simple interface that allows her to go about her activities without much concern for what is actually happening from a more technical point of view.

Central heating is a good illustration – except for adjusting the temperature once in a while and paying the energy bill, it is typically something we only notice when it does not work at all. So successful in terms of convenience and efficiency, central heating has faded from our consciousness. Now, is it really so strange that we have a hard time relating to our daily electricity consumption, consequences of our lifestyle, and effects on production, when so much has been done to hide its workings away from us? Energy-consuming products and systems, such as central heating, illustrate that it is not only a decision made at the point-of-purchase at stake here, since use, as it unfolds over time, must also be considered. At times, it seems as if the shift towards the immaterial is as much about losing touch with – as it is purported to be about the properties of – new technologies.[6]

In reality, 'buying' is already about buying into a process – it's just that such processes are primarily expressed in the shape of a product-object. This object-orientation might seem contrary to a more process-based account of sustainable design. Indeed, we somehow need to change the perception of what it means to buy something. At the same time, the object also presents an important opportunity.

Just as a product can be considered to hide aspects of a process in the form of an apparently discrete and self-contained material object, this same object can also be seen as an opportunity to express and materialize underlying processes. Turning the idea that products hide processes upside-down and inside-out, we can instead try to think of the object as a site for the expression of otherwise intangible and distant processes. This opens up for a perspective on design as unpacking, rather than 'packaging', processes.

Response

This discussion has so far presented two basic ideas about how sustainable design might be approached: first, that we can explore design based on engagement rather than problem-solving, and; secondly, that we can think of design not as packaging, but as a matter of unpacking complex issues that can perhaps become expressed only as use unfolds over time. In the research program for Static!, these two ideas are present as themes of 'working with energy as material' and 'reflective forms of use'.

energy as material

The idea of rediscovering energy as 'material' in design is grounded in our earlier work, in which we have been questioning and investigating the supposed im-materiality of information technologies in relation to the more traditional materials used in design.[7] When it comes to energy, however, issues of materiality look slightly different. Electrical power is a ubiquitous resource that we have come to take for granted, at least in the West. Indeed, electricity has even been referred to as a successful and archetypical example by technologists envisioning how information technologies will become invisibly and usefully integrated into everyday things and environments in the future.[8]

Indeed, one of Borgmann's central illustrations of how technological development seem to displace "focal things" (things that demand and reward attention and engagement), with what he refers to as "commodities", is the transformation (or even elimination) of social practices surrounding the hearth by central heating.[9] This also serves to point out, however, that the value of central heating exceeds that of the hearth in some ways – for example, its contribution to improved comfort and health is a value that is perhaps only appreciated when suddenly absent. Such different valuations of use create a further situation, in which we might have rather romanticized ideas about the qualities of log-burning fireplaces, whereas services like central heating are considered merely as neutral, technical solutions. Such perceptions reveal that the design issues related to energy use are even more complex, since somehow we need to reexamine this supposed technological neutrality to explore the range of overlapping and perhaps conflicting values.

With regard to energy consumption and its associated production and infrastructure, it seems we face a paradox. Electricity production and distribution are, on one hand, something entirely mundane that we tend to take for granted. But, on the other hand, it seems to be something quite abstract that we can hardly seem to grasp – for example, as we might try to relate to the consequences of nuclear waste, the construction of new dam or an artificial reef for an offshore wind farm.

Consider, for example, an ordinary power plug or socket. Though relatively simple in terms of spatial form, the socket is not simply a source of electricity – it is an interface to vast and complex systems acting behind. Considered in this way, energy systems disrupt normal conceptions of space just as much as information or communication technologies. As I use a computer to type this, I'm using remote information servers and communications services, but I am also using energy networks and power plants in any number of places, some perhaps not even in this country. Furthermore, time is vastly expanded beyond 'now', since the material used to produce this electricity might be as ancient as oil or might leave traces upon the future for as long as it takes for nuclear waste to decay. Now, is the 'form' of the power socket actually that simple to perceive or easy to use? This is a cautionary illustration of the deceptive power of the visual-sculptural notion of form inherited from certain art historical traditions as applied to technology design, in general, and, in particular, to sustainable development.

It seems that applying design in order to render energy more present in and through use can not only be about visualizing (for example, electricity) as such, but must also be about connecting use – and users – to the systems behind. With respect to this, aesthetics might seem to be a rather shallow perspective. But the issue, here, is not to make energy look good. Expanding our conceptions and applications of aesthetics is central because we need to explore the ways in which energy use can become more present and pressing – and thereby also something that invites reflection and engagement.[10] To get from nowhere to now and here, you just have to insert a small space _ for reflection.

reflective use

It has been suggested here that sustainable development requires us to think about consumption beyond the point-of-purchase, in terms of how use unfolds over time. Of course, decisions about what to buy (and what not to) are important, but there is a further challenge in dealing with what precedes the moment of purchase and what follows afterward. This suggests that we rethink how to work with relations between intended functionality and actual use, since the goal is not to provide a user with a solution to a given problem but to create an incitement and a space for reflection.

Although interaction design, and related fields, once grew out of a need to develop technology applications that were easy to understand and use, it is increasingly venturing into topics such as the aesthetics and use experience of technologies.[11] One reason for this expanding interest stems from a recognition that the actual use of technology often differs substantially from the use intended by its developers – a disparity that is particularly relevant if we consider what happens over time as people appropriate and incorporate technology into various practices and ecologies. Another reason is the accelerating diffusion of technology into everyday life. Since many technologies were originally intended for industrial or office contexts, use in everyday life forces developers to acknowledge values that perhaps were not originally present, such as aesthetic, emotional and social aspects.[12] Venturing beyond notions of utility and usability, this has brought new perspectives on what the use of technology is, and could be, about. Today, information technology is becoming as much about fashion as anything (or everything) else, and tech gadgets are hyped in ways quite similar to the products of other design domains. But the expanded range of values associated with technology has also opened up a space for a more critical exploration of its role.

There have long been movements at the fringes of design discourse challenging dominant ideas about functionality, utility and "good design".[13] Such movements frequently target the ideological aspects of the more mainstream (and often modernist) design approaches. Suggesting alternative forms and, significantly, alternative perspectives and intentions, proponents question not only solutions to given problems, but also parameters defining the problems themselves. Borrowing from early movements in design and conceptual art, and from a renewed (or perhaps transformed) interest in the materiality and expressiveness of technology, there is an increasing body of work reconsidering use. Instead of preoccupation with utilitarian ends or practical purposes, such work explores the complexity, indeterminacy and processes of use. 'Conceptual'- and 'critical design', in particular, apply aesthetic strategies in such an inquiry. For example, "estrangement" is applied to introduce moments of reflection, or "poetic distance", between a user and a "post-optimal" object,[14] and re-design and re-/de-construction is applied to render ordinary objects and forms of use "strangely familiar".[15]

Such design strategies and tactics have attracted considerable interest because of their potential to explore and confront values that are ordinarily hidden. Exposing such values in design can allow subsequent engagement with not only the practical functionality of a design object but with more complicated and less tangible aspects of use – thereby opening up for more reflective forms of use. This raises a further question, however, about what is meant by 'use'. The often ideological critique coming from the fringes of design can appear to verge on 'design for design's sake' when viewed from the outside – particularly if we think design should be responsive to human and worldly needs, not, or not only, an internal or disciplinary agenda.[16]

While the approach to sustainable design sketched here builds on the aesthetic strategies and conceptual implications of such critical practices, we also attempt to push certain questions concerning the relation between design and use further. One principle reason for this is that if sustainable design must engage with how use unfolds over time, we must relate to actual use to a much greater extent. To promote active engagement in everyday life, rather than the detached reflection of 'use' as experienced in an exhibition or museum, we must relate to the day-to-day interactions with everyday objects.

However, the shift from using design for critical examination of its own discourse, to using associated strategies to make people reflect and engage in everyday life, introduces a range of new issues. There is a fine line between ideological critique and persuasion – between treating the idea of use within a larger critique and articulating ideas in order to influence change among users. When explicitly mobilizing design to persuade, it is central that a user interprets the object in the way intended by its designers. Critical design, in contrast, frequently explores the indeterminacy of use and interpretation, in order to encourage resistance to rapid and easy assimilation or to achieve a certain feeling of uncertainty that, in turn, opens up a space for reflection and an expanded range of choices. Although both persuasive and critical design explicitly engage with the concepts underlying a given design, they do so on the basis of quite different perspectives on the role of use and interpretation.[17]

When it comes to working with reflective forms of use, for example, by means of critical deconstruction, we also need to relate to notions of persuasion. Of course, there will be an ideological aspect of what we do, since any challenge to existing practice can never be considered neutral – nor should it be – but neither are we presenting another solution to a given problem. We do not claim that this approach to design is about problem solving. It is not so much a matter of saying that people should do things in this way instead, as it is a matter of making people think about what they are already doing. And, so, our project might instead be considered as the proposition of alternative viewpoints. We attempt to materialize "arguments in material form", as persuasive design also aims to do[18] – but we do so by asking questions rather than providing answers.

Discussion

Here, I have argued that there is a need to revisit basic concepts in design if we are to meet the emerging challenges of sustainability. Even though such a revision means that this is, at least to some extent, a conceptual and theoretical project, we investigate conceptions of sustainability through design. In this way, a contribution to the theoretical discourse is made through critique and through counterproposals. Indeed, challenges to conceptions around the design and use of energy are not made in the form of external descriptions or retrospective analyses but as integral to the designed object as such.

This project itself is a kind of experiment with research taken through, and by means of, design.[19] We are exploring an idea that conceptually-oriented design makes it possible to work with a relation between theory and practice that is neither an empirical test of theory through practical experimentation nor a practice as applied theory – or theory as formalized practice, for that matter. Central to our investigation is how conceptual and critical design represent a shift beyond the object, toward the underlying concepts, ideas and ideologies. Historically, this can be seen in light of arguments around 'good design' or 'taste' that have been made most forcefully by means of form rather than words. Commenting on the "object as discourse", Seago and Dunne argue that "the electronic objects produced in the studio... are still 'design,' but in the sense of a 'material thesis' in which the object itself becomes a physical critique... research is interpreted as 'conceptual modeling' involving a critique of existing approaches to production/consumption communicated through highly considered artifacts".[20]

Creating and exposing relations between an intellectual discourse and a materialized object might help to bridge the gap between what we know and think about environmental issues, such as energy consumption, and what we actually do in our everyday interactions with objects. So, even if the argument presented here is, to some extent, a conceptual and theoretical project, it is founded on the belief that design can put us in touch with questions of sustainable development in sites and situations where theory cannot reach us.

Notes

1. See, for example, Shove, Watson, Hand and Ingram, *The Design of Everyday Life.*
2. For a related argument, see Allen, "Its Exercise, Under Certain Conditions," *Columbia Documents.*
3. See, for example, Dobers and Strannegård, "Design, Lifestyles and Sustainability," *Business Strategy and the Environment*; Margolin, "The Product Milieu and Social Action," in *Discovering Design*, ed. Buchanan and Margolin.
4. See, for example, Bell, "Ruins, Recycling, Smart Buildings," in *Strangely Familiar*, ed. Blauvelt.
5. See, for example, Borgmann, *Technology and the Character of Contemporary Life*; Willis, "Ontological Designing," in *Design Philosophy Papers*, ed. Willis.
6. See Redström, Redström, and Mazé, eds., *IT+Textiles*; Redström, "On Technology as Material," in *Design Philosophy Papers*, ed. Willis.
7. Fry "Rematerialisation as a Prospective Project," *Design Philosophy Papers.*
8. See, for example, Norman, *The Invisible Computer.*
9. Borgmann, *Technology.* See also Verbeek, "Devices of Engagement," *Techné.*
10. See Redström, "Persuasive Design," in *Persuasive Technology,* ed. Ijsselstijn et al.
11. See Redström, "Aesthetic Concerns," in *Pervasive Information Systems*, ed. Kourouthanassis and Giaglis.
12. See, for example, Blythe, Overbeeke, Monk, and Wright, *Funology: From Usability to Enjoyment.*
13. For further background, see Robach, ed. *Konceptdesign;* Mazé, *Occupying Time.*
14. Dunne, *Hertzian Tales.*
15. See Blauvelt, ed., *Strangely Familiar.*
16. See Mazé and Redström. "Difficult Forms" *Research Design Journal.*
17. See Redström, "Persuasive Design."
18. See Buchanan, "Declaration by Design: Rhetoric, Argument," in *Design Discourse*, ed. Margolin.
19. For background on this discussion, see Frayling, "Research in Art and Design," *Royal College of Art Research Papers.*
20. Seago and Dunne, "New Methodologies in Art and Design Research," *Design Issues.*

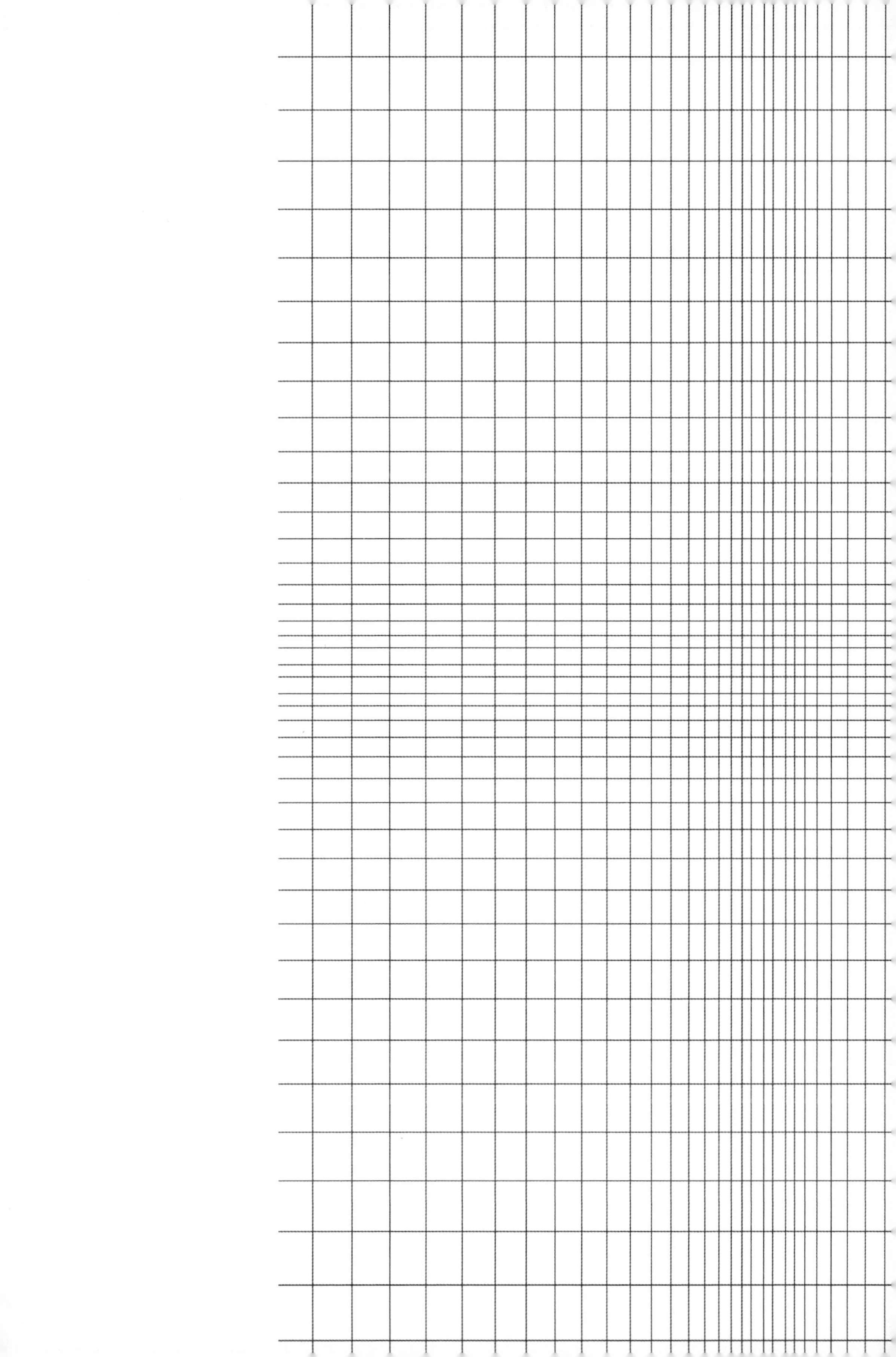

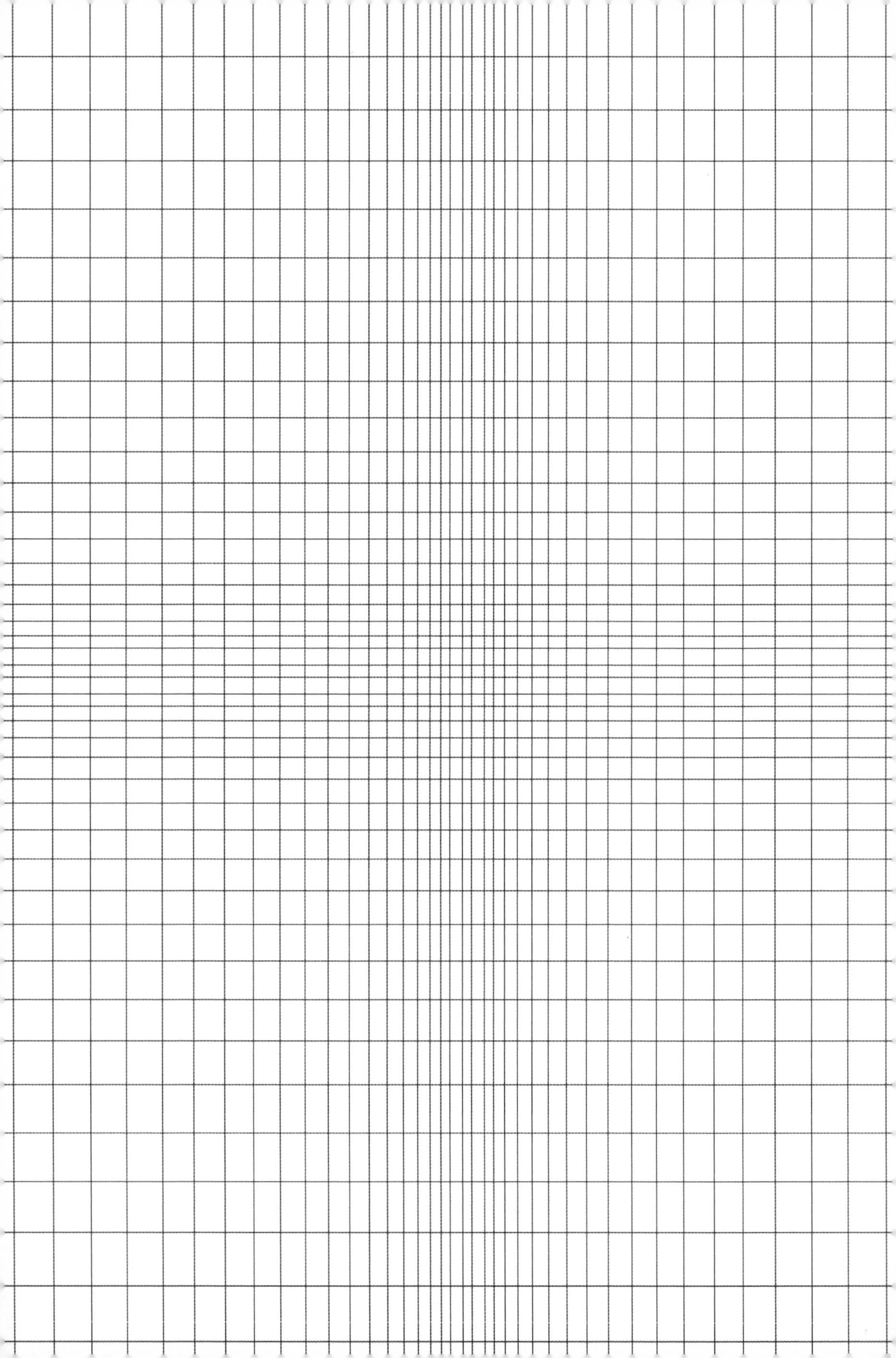

Perspectives

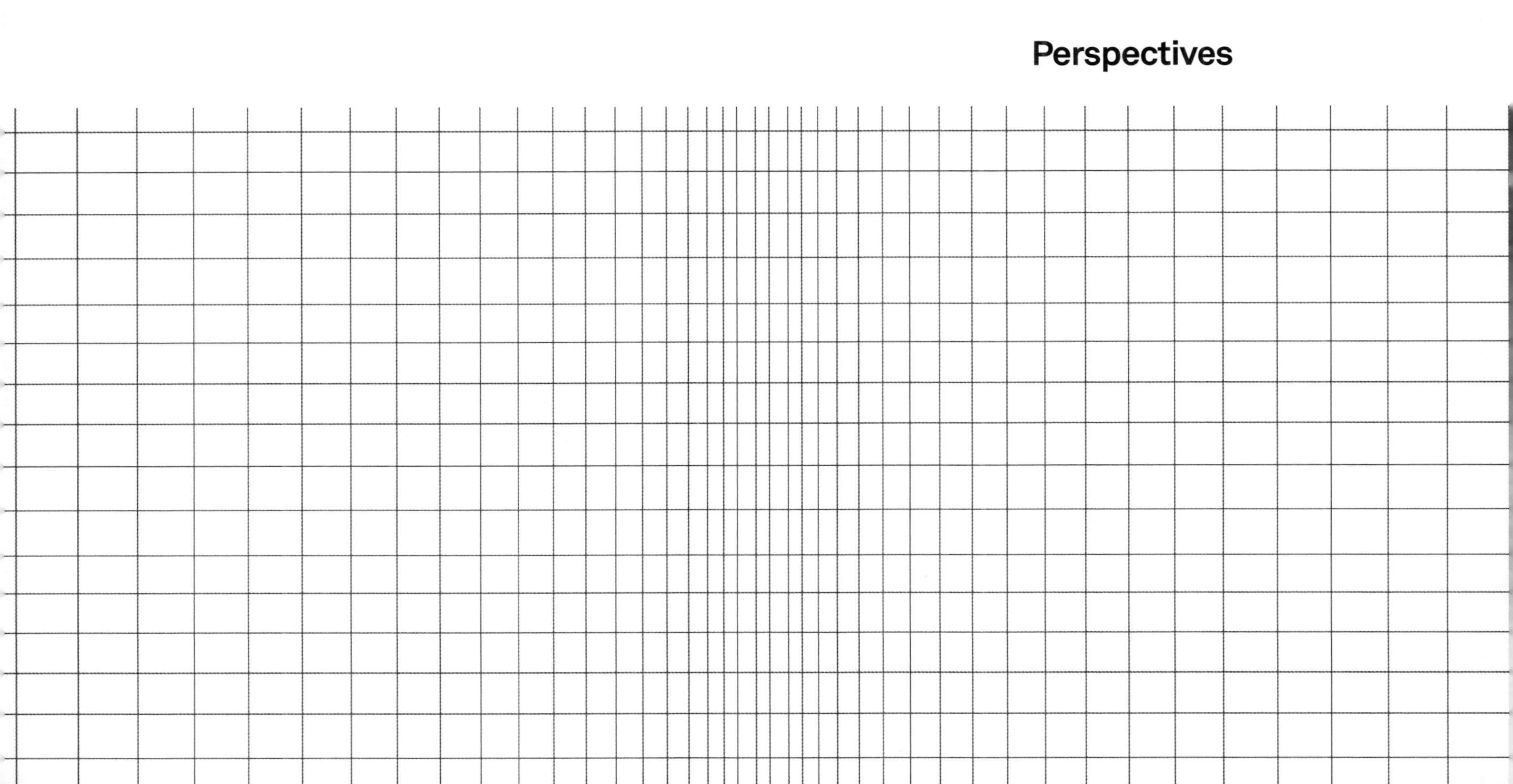

Critical practice

Ramia Mazé

In service to popular opinion and market forces, design easily becomes complicit in trends that ensure the obsolescence of what we have and create the insatiable desire for the new. Other fields have a basis for challenging such trends. Art, for example, has a long tradition of relating critically both to societal issues and its own reliance on commerce. Maker-led and hand-made, crafts practice evades dependence on mass production and mass-markets. In architecture, a vivid discourse within the discipline and the profession has long defended the need for radical forms of thinking and making. The basis for criticality in the newer fields of industrial and interaction design is less definitive. As John Thackara notes, "Because product design is thoroughly integrated in capitalist production, it is bereft of an independent critical tradition on which to base an alternative." [1] This is unfortunate for designers seeking alternatives to the mainstream – indeed, such alternatives might be essential for sustainable design today.

There are a range of emerging perspectives in contemporary design – often amended as 'conceptual' or 'critical' – that counter mainstream views of what design is and what it should be about. This is not without precedent – indeed, a heritage might be traced via radical crafts, anti-design, and critical architecture in design history. Characterizing such tendencies in a 1972 exhibition at the Museum of Modern Art in New York, Emilio Ambasz surveyed designers who, "despairing of effecting social change through design, regard their task as essentially a political one." [2] While engaging more extensively in critical theory and political action, design was seen to offer particular possibilities for "active critical participation" in larger ideological systems. Rather than 'in service' to dominant culture or capital, the act and craft of design were viewed as vehicles for contesting the status quo and for constructing alternative ethical and societal models.

Conceptual and critical design raise alternative ideas and tactics. Drawing on conceptual art, conceptual design engages craft and materials but does so in order to shift focus from the maker and the object to the concept behind. For example, high and low materials, precious substances, ready-mades, technology and trash, may be combined to expose assumptions of 'taste' and 'good' design – even material scarcity may speak to ethics and (over-)consumption. Critical design applies techniques of decontextualization and defamiliarization from modern aesthetics and critical theory. For example, alterations in form or function that disrupt 'utility' and 'usability' might make unthinking assimilation impossible, requiring people to reflect upon the norms embedded in things. A variety of such tendencies towards what might be generally called 'critical practice' are amassing an increasing number of examples, theoretical depth and public exposure.[3]

The products of critical practice do not aim to solve or resolve the problems in focus in mainstream professional practice – instead, they expose underlying assumptions and ideologies. Indeed, critical practice is less concerned with problem-solving than 'problem-finding' (to borrow a term from anti-design) within disciplinary and societal discourse. Sharing common ground with research into the history and theory of design, critical practice locates questions and makes critiques. However, design practice remains central, including the creation of objects as a 'material thesis' – a materialized or physical form of discourse.[4] Creating new relations between the theoretical and practical aspects of design, critical practice opens up for new forms of practice-based design research as well as new forms of ideologically-oriented design practice.

Besides 'problem-finding' for it's own sake, critical practice is concerned with opening up design ideas for wider reflection – as well as criticism and debate. In exposing alternatives to mainstream design, through design products, critical practice can be seen as a way of opening up questions to consumers and other audiences. Indeed, design materials and form can expose certain problematics in ways that make them more accessible to understanding – and to change. However, in order to operate from within design while simultaneously raising a critique, critical practice also requires alternatives to mainstream modes of production and consumption.

For example, critical practice seeks other roles for producers and modes of production. The designer might seem to shift to the background, in order to shift a concept or idea to the front. Accomplishing such a shift, however, requires significant design knowledge and skill, for example to account for material history, craft techniques, context of use and consumer perception. Familiar materials, for example, are subtly repurposed – plywood might be applied as a neutral material to encourage the reading of an object as a concept model or, its seams and faults exposed, might comment upon the repressive standardization of modernism. Forms might be deliberately underdesigned or overloaded with symbolism – techniques of applied ambiguity. Understood in relation to history and culture, materials and form must be carefully selected and crafted to achieve the expression of alternative ideas.

Further, resulting objects are intended to produce other forms of consumption. Aaron Betsky argues that such "designers prefer to create something abstract and alien, precisely because it denies their contributions and opens the object, image, or space to multiple interpretations." [5] Strangely familiar things give pause and require interpretation. Consumption of critical objects, in fact, becomes a form of 'active critical participation'. This might place the consumer as a critical and informed reader, consciously deconstructing the ideologies inscribed by the designer into the form of an object, or elicit an anarchy of subjective interpretations that seem to deny the possibility of any singular meaning. Instead of consumption on the basis of mass production and mass markets, alternative modes of production might range from small-scale custom manufacture to display in art museums and journals. Carefully situating consumption and consumers, critical practice develops alternative modes for its 'ideological' production.

Static! investigates this tension – between critical objects and critical subjects. Drawing on design history and theory, as well as a range of design and technical practices, we designed, or redesigned, things with an alternative aesthetics of energy and energy use. While not all examples relate to the design tactics of critical practice, some delve into dys/functionality or (un)ease-in-use of electrical things. Each experiments in different ways with altering sensory perception and courses of action in order to redirect the focus of attention to energy in everyday things and mundane interactions. Collectively, they represent variations on the insertion of a 'critical distance' between design convention and normal consumption in order to expose the (inter)dependency between energy and products and actions.

In Static!, we move beyond 'problem-finding' to give form to energy issues that might promote self-reflection in use. Indeed, this concern for use breaks with the theoretical dialectics of discourse around critical practice. As Anthony Dunne argues, "To provide conditions where users can be provoked to reflect on their everyday experience of electronic objects, it is necessary to go beyond forms of estrangement grounded in the visual and instead explore the 'aesthetics of use' grounded in functionality, turning to a form of strangeness that lends the object a purposefulness." [6] Repurposing utilitarian objects and mundane situations, energy use is taken conceptually and literally in Static! An alternative aesthetics of energy use is, on one hand, removed from actual use and put on exhibition for debate among designers, stakeholders, and the public. On the other hand, alternatives are (provisionally) reinserted into households to see how people might rediscover their own relations to energy in and through their interactions with the objects.

Critical practice circumscribes one set of theoretical and practical strategies in and around questions of ideology, criticality, and accountability in design history and contemporary practice. Such tendencies operate to distance us from conventional notions of production and consumption – in design, such radical perspectives and dissenting examples essential to the development of a disciplinary discourse. Further, criticality may not only be about greater retrospection or reflection on our own practice, but also about materializing the status quo – and alternatives – in order to open up for wider debate. Outside design, critical practice may operate to stimulate reflection upon one's own relations to consumption and perhaps even change of attitude and behavior.

Notes

1. Thackara, "Beyond the Object," in *Design after Modernism*, ed. Thackara, 22.
2. Ambasz, *Italy: The New Domestic Landscape*, overleaf.
3. See also, Mazé, *Occupying Time.*
4. See Seago and Dunne, "New Methodologies in Art and Design Research," *Design Issues.*
5. Betsky, "The Strangeness of the Familiar," in *Strangely Familiar*, ed. Blauvelt, 42.
6. Dunne, *Hertzian Tales*, 42

People in design
by Sara Ilstedt Hjelm

Today, there is a near consensus in design that products should be developed with and for end users. Large corporations talk about user-driven design, participatory design involves end users directly in the design process, and marketing studies consumption behaviors and lifestyles. This, however, has not always been the case.

One of the most influential design academies in modern times is the Bauhaus, a German school and design think-tank active between 1919 and 1933.[1] In the Bauhaus manifesto (and numerous other texts) written by founder Walter Gropius, there is nothing about users. Not that the term 'user' is very attractive – nowadays we prefer to say 'people' or to be more specific about the type and context of the people referred to, for example, 'nurses at a children's hospital' or 'single males between 22 and 35'. Gropius does not denote the people who are supposed to use the products of the Bauhaus with a word. Another notable shortcoming in his texts is of women – students are 'him' or 'his' and the professors are male. Design was evidently to be done by a few (men) who knew best, for the benefit of the rest of society. Luckily, this has changed. People, users, women, and other human majorities are now at the center of design agendas.

In other ways, however, the Bauhaus is an example of influential design theory and practice addressing important issues in society. The main challenge for Gropius and his colleagues at that time was mass production. 'The machine' was said to have ruined craft, disrupted aesthetics and brought about a division of labor. Gropius argued that the machine was also an opportunity to free mankind from tedious work. To address both the opportunities and threats of new technology, Gropius argued, creative people from different disciplines needed to come together and create a new aesthetics. Today, we have come to terms with machine-made products and with the fact that industrially-produced things can have aesthetic value. But mass production has brought about other problems that urgently need our creative attention. The environment is such an issue.

In order to prove that machine-produced objects could have high aesthetic quality, the Bauhaus formed design prototypes. These were usually handmade in limited quantities, but they looked like they could have been mass-produced. For example, the steel furniture by Marcel Breuer was cumbersome to make by hand but effectively conveyed a new aesthetic paradigm. Thus, the Bauhaus communicated the idea of a new agenda where designers cooperated with industry to create products for a mass market. But, in reality, little of Bauhaus design was mass-produced – only a few lamps, some wallpapers and textile, designed, ironically, by the few female students reluctantly admitted to the school. The work from the Bauhaus was always striking in appearance, but more conceptual than practical. For the general public in Germany at the time, it must have been regarded with the utmost skepticism.

But the impact of the Bauhaus on the design community afterwards cannot be underestimated. Steel furniture, white cubist buildings and abstract patterns will always be connected to Bauhaus, as well as educational models and theories. Therefore, I would argue that the profound contribution of the Bauhaus was on a conceptual level. They criticized existing aesthetic ideas and argued their point with both texts and design prototypes, the main function of which was to be discussed rather than produced. Today, such a design agenda might be labeled conceptual or critical design. If the goal is merely to make attractive products, design should be comfortable and reaffirming, rather than questioning. But, if design is to break norms and introduce new ways of thinking, to some extent it needs to be uncomfortable and to place itself outside the current paradigm.

The relation between critical design and user-centered design is not without friction, since each stems from a different tradition. User-centered design developed out of a need to adapt new, complex technology to people. Traditionally, user-centered design was characterized by studies carried out by sociologists and anthropologists or human factors specialists, the former observing people in their natural surroundings and the latter in an experimental setting. Results of such studies were handed over to designers in the form of written reports, which may or may not actually have been used. During recent decades, however, designers have increasingly started to take an active part in user studies.

This has resulted in a number of new methods, since designers may not be satisfied with merely 'observing people' and want to involve people more in the process. For example, participatory design (originally called cooperative design), developed in Scandinavia in the early 1980s to involve workers directly in workplace design. The 'cultural probes' method,[2] another example, was developed by researchers (from psychology and design) who wanted input from people without only taking the role of 'studying' them. Another driving force behind user-centered methods has been the idea that products should last longer and therefore need to be emotionally sustainable. To reduce consumption without compromising quality of life, we need ways of developing objects that create a deeper and more meaningful relation to their owners.[3] In order to do that, designers need to know who they are designing for. A variety of formats have also been developed for involving users in design, through interviews, focus groups, surveys and other means. Today, we see methods spanning from anthropometric studies of physiological and psychological factors, to contextual inquiry into people within their daily contexts, to participatory design.[4]

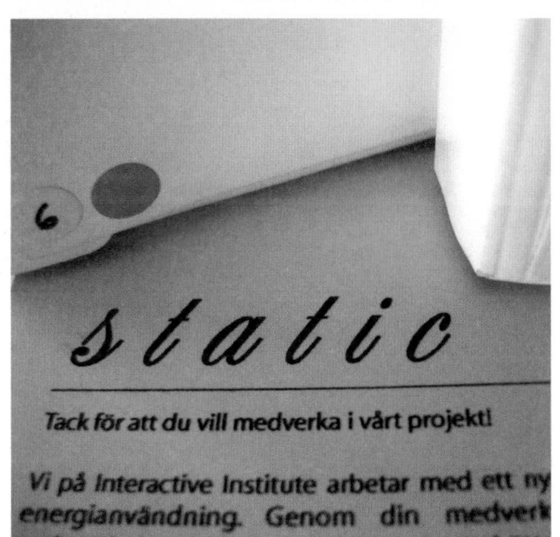
static
Tack för att du vill medverka i vårt projekt!
Vi på Interactive Institute arbetar med ett
energianvändning. Genom din
undersökningar kan du hjälpa oss att

datum: tid: plats:
Beskriv vad du fotograferade
Vad kommer dess energi ifrån?

TVÄTI
TRANSISTOR INTERCOM
Svenska sjöfågla

In Static!, our work was people-centered for several reasons – for inspiration, for new ideas, to gather facts, to give a context and to create emotional sustainability. For example, we used methods including cultural probes and visits with interviews in private homes. Distributed to ten households, each cultural probe consisted of a disposable camera and a diary, which encouraged participants to reflect upon their encounters with energy throughout the day. In another group, each team member was given the task to visit a home, conduct an interview and take photos. These activities resulted in materials that were then analyzed collectively. Some design ideas were directly inspired by the studies. For example, several respondents mentioned that they sometimes left a lamp on in an empty house, because "it also generated some heat." Prompting us to think further about the connection between heat and light, this was a starting point for 'The Element'. Other materials were less directly connected to specific designs produced, but nevertheless built up a context for our work.

The design process contains a mixture of rational analyses and intuitive choices, based on deep practical and aesthetic knowledge. Since the early 1970s, researchers have talked about a certain type of "design thinking" [5] that is common to design disciplines. Stolterman describes design as "an activity where thinking is very closely connected to the hand, with the physical world, with the material and the complex reality".[6] A designer cannot base his or her work on the kinds of descriptions privileged in science, just as you cannot learn to ride a bike from a book only. Knowledge needs to be seen in context, in situations and as a physical event. This is why, in Static!, although there are studies that cover consumer behavior regarding energy use, it was important for us to base our work on first-hand experience. Collecting such impressions and experiences of use and, from there, proceeding to create a unity – a product – is fundamental to the synthesizing work within the design process.

Many have also pointed out the tacit knowledge that several professional fields are based on. Molander describes the knowledge in action as a form of "embodied knowledge".[7] A hospital nurse can sense that a patient soon will pass away but cannot explain why. Likewise, the designer can sense a solution but might not be able to verbalize it. Sketching or prototyping in the workshop can help to externalize an idea. Knowledge lives in the meeting between the hand and the material. Indeed, in the Bauhaus, every student had substantial practical training in different material and workshop techniques. The first year of study was dedicated to a foundation course in which students tried out various crafts. It was out of this deep insight and practice-based knowledge that they believed a new aesthetic for the future would arise.

In Static!, we argue that energy can be regarded as a material in the same way as wood, metal or textile. In order to visualize and become aware of energy in our homes, we need to apply the same care and knowledge to energy as to the other materials as we design products. Therefore one important part of our work was to experiment with different approaches to energy as a material, as well as 'using' energy to design with. To conclude, in Static!, we tried to merge a critical and reflective stance with one oriented towards people and use, creating objects to establish a sustainable emotional connection and that inspire reflection.

Notes

1. For further background see, for example, Fiedler, Feierabend, and Ackermann, eds., *Bauhaus (Design)*.
2. See Gaver, Dunne, and Pacenti, "Cultural Probes," *Interactions*.
3. See Chapman, *Emotionally Durable Design*.
4. For a survey of various methods and perspectives, see Laurel, *Design Research*; Moggridge, *Designing Interactions*.
5. See, for example, Cross, ed., *Developments in Design Methodology*.
6. Stolterman, "Designtänkande," in *Under Ytan*, ed. Ilstedt Hjelm.
7. See Molander, *Kunskap i Handling*.

TYP XS 315
15 A 380 V~

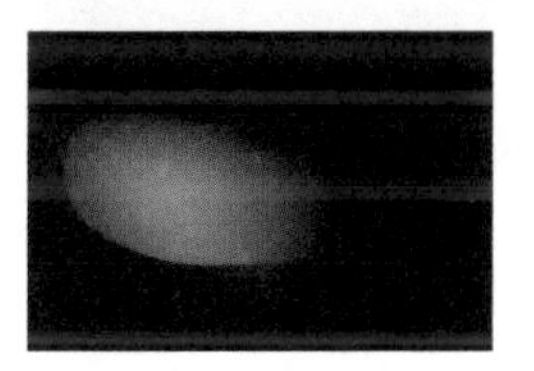

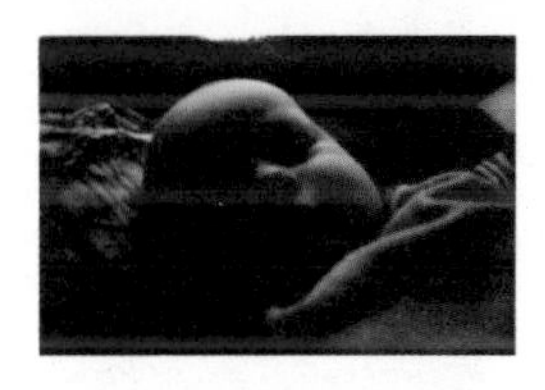

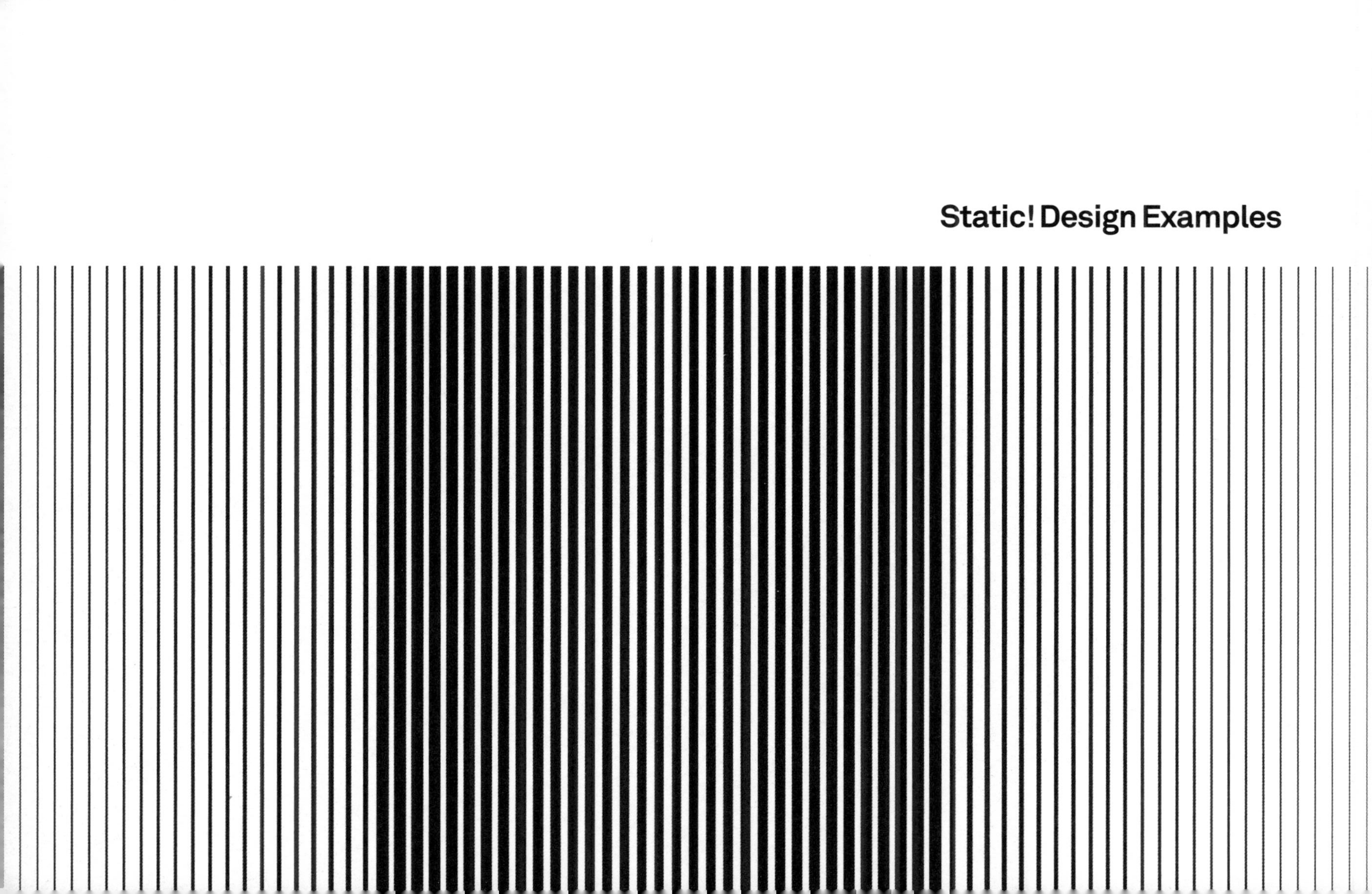
Static! Design Examples

The Element

Especially in northern countries such as Sweden, light plays an important role in the home. Just consider the symbolic as well as functional role of candlelight and the hearth in family life! In a series of interviews in local homes at the start of Static!, we investigated the emotional and cultural significance of energy in the household. Drawing on the impressions gathered, the 'Element' explored the materiality of energy – and how different material forms might affect people's engagement with the energy that they already use everyday.

In the interviews, we discovered a range of relationships and rituals around lamps and other sources of light. While lamps seem to occupy a significant amount of attention as well as space in the home, other electric products tended to be seen as solely functional objects – hidden away or disguised. For example, while people typically seemed to buy the most inexpensive and efficient radiator, lamps often had meaningful stories attached to them and were often named as favorite household products.

While this may not seem surprising, this highlights something significant from the point of view of sustainability. While radiators consume a much greater amount of energy, we seem to place a higher value on lamps. In order to focus on developing products that better visualize or materialize the energy consumed, we reverse the order of importance placed on lamps and radiators in the Element – thereby 'casting light' on radiators as powerful forms in the local and wider environment.

The Element is a radiator constructed from thirty-five ordinary light bulbs. The array of sixty-watt bulbs generates the same amount of heat as a conventional electric radiator (approximately 2000 watts). The case that secures the bulbs in place also contains sensors and a microprocessor that gauge the optimal and the actual temperature in the room and control the intensity of the bulbs accordingly. A control wheel is located on the right side of the case, which allows users to input their temperature preferences and view temperature changes.

When the appliance is turned on, it slowly starts to glow, growing increasingly brighter until the temperature in the room rises to the value set by the control wheel. If a door or window is opened, or as other factors change the temperature in the room, this is balanced as more heat – and light – are emitted. The climate of the room, thus, is both visualized and adjusted by the vibrant presence of the Element. The constant visual and temperature modulation of the Element reflects the continual negotiation between heat and light in maintaining a constant state.

Blurring the product semantics of a lamps and a radiator, the Element literally combines the two in a single form. The design was inspired from the household interviews – in particular, an observation that lamps were left turned on because, as one homeowner expressed it, "lamps also generate heat so it is not a waste at all." By combining the properties of a lamp and a radiator, the Element makes use of light to visualize heat and, vice versa, makes use of the 'waste' heat of light bulbs.

Conceptually, the project refers to the law of energy conservation in physics, which we interpret here as an energy recycling concept. The law states that energy can never be destroyed but can only change form. In the Element, electricity used by the device is transformed into 70% heat and 30% light. As the light is absorbed by surrounding objects in the domestic context, it is eventually turned into heat as well. The element will, ultimately, produce 100% heat, as well as 30% light, using the electricity to simultaneously fulfill two previously separate functions.

Interpretations of the Element were explored in a session with ten potential users. As the participants tried out and experienced the working prototype, they expressed strong emotional responses to the product's aesthetics. In terms of product semantics, the first reading of the object was often as a lamp, though further exploration indicated increased engagement. As participants began to understand the product as a radiator, they began to reflect more deeply on the relation – and balancing act – between light and heat, and form and function, of such products

Project and evaluation team: Anton Gustafsson, Magnus Gyllenswärd

Appearing-Pattern Wallpaper

To develop an alternative perspective on domestic products, we collaborated with the celebrated design group Front. Then only recently graduated, they had created a series of works exploring 'design by nature', in which physical forces, natural processes and animal behaviors were determining factors in design. For example, an explosion in the ground and dog tracks in the snow generated the form of a chair and a vase, and patterns in the surface of a wallpaper and a table were created by rats and insects eating the material. They joined into our investigation of 'energy as a material'.

One of their early projects was the 'Appearing-Pattern Wallpaper', which exposed the aesthetics of a natural energy source – sunlight. Sun is often viewed as having an undesirable effect, since its ultraviolet spectrum can cause bleaching or decay in paper, textiles and other materials. Typically, in design, materials science and product manufacture, much effort is made to prevent sun damage over time – however, in design history and the art market, there is another appreciation for the value of rust, patina, and other telltale signs of time. The Appearing-Pattern Wallpaper, accordingly, makes a virtue out of exposure to sunlight. Instead of decreasing in value as it deteriorates, the Wallpaper becomes more beautiful the longer it is used.

At first, the Wallpaper appears to be an ordinary monochromatic paper but, over time, a pattern begins to emerge where it is exposed to the sun. The Wallpaper is fabricated using two pigmentation techniques – the pattern is printed in a conventional UV-resistant pigment, overlaid with a pigment that deteriorates over time. During the lifespan of the Wallpaper, the latter will bleach where the sun shines on it, slowly exposing the hidden pattern. Energy – in the form of sunlight – is thus made visible and aesthetic. The (de)composition of the Wallpaper pattern is contingent on nature, since the light and movement of the sun will vary on a daily, seasonal and yearly basis. Further, domestic life also comes into play, since the window treatment and furniture placement, which might also change seasonally or over the years, frames the contact between the sun and the walls. Instead of design objectifying a permanent or timeless aesthetic, the Appearing-Pattern Wallpaper continues to evolve, organically, with use over time.

This idea – design by energy – is present in several of Front's concepts. In another, the 'Heat-Sensitive Lamp', the shade of an ordinary table lamp is made of a material that reacts to heat. The choice of light bulb and the act of switching on the Lamp converge to instantly give the lampshade its form. Energy has an active role in the formation of the Lamp and the Wallpaper – electrical wattage gives form to the Lamp instantaneously and sunlight patterns the Wallpaper according to the slow cycles of nature. Indeed, as it was produced in a limited quantity for installation in a simulated domestic environment for an exhibition, it showed dramatic, and delightful, changes over the course of a Swedish summer filled with long, light days.

This extends energy as a material – indeed, energy might even be considered as a designer of sorts. Further, this extends the consequences of consumption – the choice of light bulb or wallpaper is revealed as more than a one-off decision, but a choice with effects that can be experienced tangibly and in the long term. Beyond the act of design, and after the point of purchase, formation continues – by nature and by use – throughout the lifespan of the product.

Project team: Sofia Lagerkvist, Charlotte von der Lancken, Anna Lindgren, Katja Sävström (Front Design) in collaboration with Spets

Energy Curtain

The Energy Curtain embodies a trade-off that we make all the time, though perhaps without thinking about it. Consume or conserve – each act of opening or closing the Curtain is a local and tangible choice about energy use.

Like any ordinary curtain, the Energy Curtain has two sides. In this case, the one facing to the outside has solar cells, and the inside one is woven with fiberoptics. Sun shining in during the day is collected by the solar cells and stored in batteries. When night falls, and insufficient light is sensed, battery-powered LEDs distribute light along optical fibers woven into the textile pattern. Thus, the Curtain collects energy as the sun shines – saving and storing energy during the day to light up the room once the sun has set.

The Energy Curtain reinterprets our familiar relation to curtains as a means of controlling the light in a room – but with a conceptual twist. The Curtain must be drawn shut to collect light, and the amount and duration that it is drawn during the day determines how much light is collected for the night. Users must make a choice – whether to open the curtain and enjoy the daylight, or to close it and save energy for later. Thus, even the mundane act of opening or closing the Curtain embodies the trade-off between consuming or conserving energy. Each and every day, it requires that its user reflect and act upon this trade-off – literally taking the cyclical transformation of energy into their own hands.

Two versions of the Energy Curtain have been implemented. The first version was crafted to amplify the conceptual aspect. In the form of horizontally-folding Roman blinds, the two sides of this prototype are isolated with an additional lining of black-out material. Thus, the choice to consume/conserve – embodied in the act of raising or lowering the curtain – is exaggerated as the room turns suddenly turned from dark to light, or vice versa. While the appearance of this prototype may appear quite ordinary in the daytime, even if drawn to shut out the daylight, it slowly gains a dynamic and glowing aesthetic pattern in the evening. Indeed, the room might be brightest at night as the Curtain slowly comes to life, thus effecting a rather dramatic experience of sunlight delayed and deferred.

The concept behind the Energy Curtain was inspired by a response given in our study using the cultural probes method. The respondent recalled one of our earliest learned understandings of solar and other forms of energy – as a cycle of transformation dependant on daily and seasonal patterns. The Curtain is designed to draw attention to natural cycles, as well as to amplify the interdependence between patterns in nature and in human activity.

This is articulated in the material form and form of interaction with the Energy Curtain. The Curtain is crafted with energy as a basic material – and use is required to realize its aesthetic and functional form. The design requires that action be taken during the day to enjoy the effects at night. This action is distilled into each discrete act of raising or lowering the Curtain – amplified by the design to stimulate reflection on the choices implicit even in this mundane action. Furthermore, since this interaction with the Curtain must be made each and every day, this reflection on action would occur again and again. Beyond an initial reaction to the conceptual design, repeated interactions might potentially support an ongoing relationship with an object sustained by one's own energy behaviors.

To further investigate how people might develop a relationship with such a conceptual design – and perhaps with their own energy behaviors – the prototype was part of a multi-household 'domestication' study. Since the study took place in Finland over several (winter) months, the functionality of the Curtain was limited by rather extreme (dark) seasonal conditions. Perhaps because there was very little sunlight to collect and, thus, little night-light to enjoy, unexpected conversations were sparked. Local and personal characteristics of energy came to the fore, and there was a discussion of differences between daylight, electric light and the Curtain's light. Surprising side-effects also emerged – one family even powered the Curtain using a portable lamp! This (mis-)use demonstrated not only a grasp of the concept but an act of appropriation for other purposes entirely – providing a powerful example of a product relationship emerging through a process of domestication.

Second versions of the Curtain were more general purpose and for a wide audience. A series of vertical blinds were designed and produced in close collaboration with the textile manufacturing company Ludvig Svensson. Building on their interest in incorporating new information and energy technologies, the design was both technically and strategically in line with their established product line and business development. Each panel was based on the aesthetics and materials of their existing line, including many of the machine specifications, weaving techniques and finishing processes of their currently operating industrial production. These versions have led to discussions about the prospects for commercialization involving other stakeholders in the household goods and high-tech sectors.

Project team: Anders Ernevi, Margot Jacobs, Ramia Mazé, Carolin Müller, Johan Redström. with Linda Worbin (Swedish School of Textiles at the University College of Borås); Evaluation team: Sara Routarinne (University of Art and Design, Helsinki)

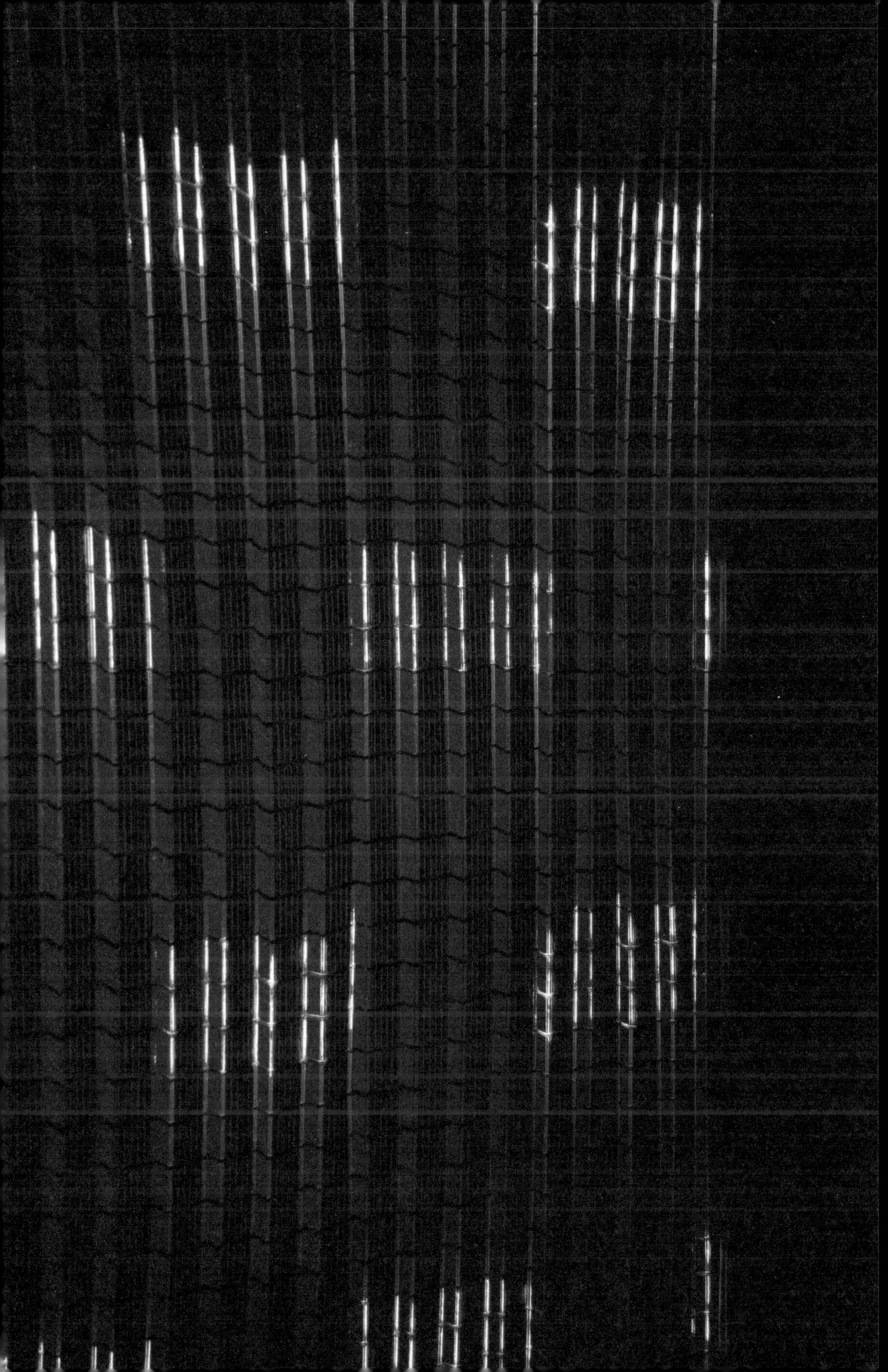

Erratic Appliances

In limited or unstable electrical systems, our actions may have rather dramatic and unexpected effects. Plugging in too many things at once can cause a circuit to blow or sparks to fly! Or, turning on the washing machine can mean that the radio and the blender will not work properly. This example draws inspiration from such familiar energy (mis)behaviors to remind us of the interdependence between global systems and local choices.

The behavior of an electric or electronic appliance depends on the overall energy consumption of the system it is connected to, for example the electrical system within a house or apartment building. Exceeding the limits of such a system introduces a risk that the things we have come to rely upon might stop working properly. While the results of our choices often seem to have unforeseen environmental consequences only long after, the 'Erratic Appliances' experiments with some more immediate forms of feedback, creating a tighter and more local relation between actions and reactions.

As a starting point, we created a series of design sketches illustrating how certain limitations or fluctuations in a local energy system might interfere with the basic functionality of ordinary household appliances. We looked specifically at instances in which electrical interference directly affected the primary function, which would thus reveal energy as an essential material determining the normal appearance or behavior of an appliance. For example, an erratic behavior of a radio might imply that it tunes out, forcing us to listen to something else. Treating energy as material integral to both functionality and aesthetics required us to delve rather precisely into how the appliance works. On this basis, even seemingly minor interventions in dependant relations between construction and behavior can significantly affect how it is experienced in use.

Based on our initial sketches, a working prototype of an 'Erratic Radio' was developed. The primary function of a radio is to tune into a specific frequency. The selected signal is decoded, amplified, and played through the speaker – thus, its functionality depends on energy as a material in several different ways. The Radio was realized by hacking into the frequency selection components. It contains an additional receiver that listens to frequencies around 50Hz – the range emitted by electric appliances. The ordinary tuner is made dependant upon this additional one, such that changes in the electrical fields of nearby appliances will cause the Radio to deviate from its normal operation. A microcontroller affects the voltage applied to the extra capacitor, changing its value along with the total value of the tuning capacitor, causing the channel selection to drift. Excessive electricity use nearby causes the Radio to drift out of tune.

The Erratic Radio is a sort of a materialization (through sound) of local energy consumption. While a normal radio is ever-present, seemingly invisible, in the background, this one attracts attention. Indeed, ordinary use of the Radio requires a change in our own behavior in use – a heightened awareness and careful negotiation of the electrical things used in the vicinity at the same time. A study of how the Radio was interpreted initially – and 'domesticated' in use over time – was made as part of an extended evaluation of several households in Finland. The study revealed much initial curiosity about the Radio, though the relation of its behavior to energy consumption did not always seem self-evident. Over time, the small size and idiosyncratic behavior of the appliance led it to be marginalized in some households, though a certain mystique seemed to create an attachment that meant they did not want to give it up.

While it is not necessarily the case that the use – or misuse – of electricity must be at the expense of usability and functionality, this example interferes with ease-of-use in order to shift energy use to the foreground of attention. The 'Erratic Appliances' are crafted to embody some potential consequences of over-consumption, such as increased indeterminacy and risk, implying the need to make choices about which things to use and when, not only on a global but on a local basis. As a rather humorous take on strangely familiar products and their behaviors, the aim is to stimulate reflection on certain larger questions – the resolution of which are up to us, in use.

Project team: Anders Ernevi, Samuel Palm, Johan Redström; Evaluation team: Sara Routarinne (University of Art and Design, Helsinki)

Power-Aware Cord

Power cables are used on a daily basis as a means of transporting energy to the electrical appliances we use. Televisions, irons, vacuum cleaners, and lamps... but how much energy do they consume? The 'Power-Aware Cord' visualizes, rather than conceals, properties of the electrical current.

The Power-Aware Cord is a re-design of a common electrical extension strip that makes visible the electricity flowing through it to electrical appliances plugged into its extension sockets. The cable construction is augmented with dynamic effects that represent the electric current as glowing and moving light. Plugging or unplugging electrical devices into the sockets results in direct visual feedback, which gives a feeling of both seeing and interacting with electricity.

The Cord may be used as a sort of 'tool' for people to rediscover energy in their homes as well as an ambient 'display' to see energy consumption at a glance at any given time. For instance, the effects of changing the volume on stereo equipment becomes immediately and dramatically apparent – as do appliances that are silently stealing electricity while on standby. With the Cord, people can learn about the energy consumption of their various home appliances, as well as the effects of operating and interacting with them.

Representing electricity with light, rather than with numbers or graphs, turns an ordinary object into an aesthetic – as well as informative – display in the home. Even placed on the periphery, it may be viewed from anywhere in a room – the energy flowing through the cord is visible and constantly available, even at a distance. Thus, the Cord also acts as a sort of ambient display of current energy use. As a tool or display, the Cord can fit in, or stand out, depending on where it is placed and how it is used in the domestic context.

The Cord has been produced as a limited series of working prototypes. Each prototype is the size and shape of an ordinary power strip, with the addition of voltage-measuring electronics and electroluminescent wires. Bound together with ordinary copper wire for electric conduction, the cable is constructed with three additional electroluminescent wires. These contain a semiconductive layer that glows with a blue-green light when an alternating current is introduced. The cable appears to be white when not powered and starts to glow blue when current flows through – since the composite cable changes not only from unlit to lit, but also changes in color, the effect of introducing electricity is dramatic. Twisted together to improve the flexibility of the cable, the cable construction facilitates an additional visual effect of motion, since each wire can be powered in turn to achieve an animated effect. The combination of the state, color and motion of light in the cable make it possible to differentiate a wide range of visual intensities, which represent variations in voltage ranging from zero to 2000 watts. The Power-Aware Cord construction is protected with a layer of transparent silicone.

Prototypes of the Power-Aware Cord have been deployed in several evaluations with potential users. Such evaluation situations are a basis for us to observe and test people's perception of the design – and, ultimately, to better understand its potential for changing their awareness of energy consumption and their behaviors in use over time. An early evaluation investigated perceptions of light as electricity – different light effects were tested, including static intensity, pulsating intensity, and moving intensity. Findings of the study demonstrated a rapid grasp of the Cord's functionality. Further, various uses were suggested by participants in the study – for example, one woman explained how she would use it to teach her children about electricity.

More recently, several prototypes have been deployed in an in-depth study as part of an environmental initiative in Stockholm. The study, conducted by a doctoral student in social science, examined the 'domestication' of the Cord in terms of change in perception in five households over three months. While the familiar appearance and function of the object facilitated its introduction into the home, the Cord did prove to challenge to expectation. Its deviation from the normally passive and unexpressive role of an ordinary power strip created surprise and much initial experimentation. Deeper relationships to the object emerged over longer periods of time – for example, one participant related to it as a "guilty conscience" nagging at the corner of the eye as a reminder of appliances "eating your money". For others, its ambient and dynamic behavior evoked perceptions of a living or organic force. By reinforcing a relation to energy as a natural resource, this interpretation led to a positive and ongoing relationship with the object, and to a change in self-awareness and attitude toward energy use.

The patented Power-Aware Cord is set for launch in the domestic appliance market, and further initiatives are underway to implement commercial and industrial applications.

Project team: Anton Gustafsson, Magnus Gyllenswärd; Evaluation team: Erica Löfstrom (Linköping University)

Disappearing-Pattern Tiles

Heat is a form of energy that is often taken for granted, invisibly accompanying lamps, emanating from radiators, and escaping from kettles. Hot water, in particular, has implications for sustainability. The product of two systems in the home – electricity and water – hot water is a doubly valuable resource. Its value can be measured, of course, in terms of economic or environmental cost. But hot water is also an example of the existential value of energy use, since it has an important role in our health and comfort at home. But how – or, when and where – might we reflect upon these values?

The domestic bathroom is already an intersection of different values: taste and decorum, hygiene and efficiency, and – considering energy – personal as well as ecological well-being. The 'Disappearing-Pattern Tiles' intervene into this situation. The Tiles react to heat, changing appearance with exposure to hot water or steam. Installed on surfaces in the bathroom, the duration of a shower or waste of hot water is reflected in the Tile pattern. With excessive heat, the decorative pattern of the Tiles fades. To maintain or regain the original appearance of the room, users have to mind their energy consumption. Even taking a shower, thus, involves the performance of a balancing act between the existential and environmental values overlapping in the use of hot water.

This example expresses our intention to relate to energy through objects that are already familiar in meaningful. The cultural probes and household interviews were methods employed early in Static! to excavate relationships and rituals involving ordinary objects in domestic life. Findings collected through these methods revealed a range of artifacts and behaviors. For example, photographic documentation, diary entries, and responses to questions revealed the private meanings of curtains, lamps, wallpaper, electrical sockets, plumbing fixtures and consumer appliances. Personal actions and interactions included drawing the curtains, turning things on and off, taking a shower and paying the bills. These presented us with a series of sites and situations where energy could be materialized, the choices and the values involved in use exposed.

Mundane and habitual, showering is not typically the focus of attention, in design or in use. Often occurring at the transition of one activity to another, or one day to the next, the experience of showering seems to remain at the edges of our attention. The Tiles, installed on one or more walls in the bathroom, amplify both the space and the time of showering. The presence of heat becomes visible on surrounding surfaces, experienced as a subtle but immersive ambiance. The dynamics of the (dis)appearing pattern expose the time passing during a shower and small changes in daily routine. This experience of showering, and the existential and environmental values involved, are thus brought to the forefront of attention – and reflection.

Project team: Sofia Lagerkvist, Charlotte von der Lancken, Anna Lindgren, Katja Sävström (Front Design)

Flower Lamp

Household lamps typically have very basic functionality with respect to energy, expressed in lit states of 'on' or 'off'. Besides turning electricity into visible light, as any lamp would do, the 'Flower Lamp' builds on an increasingly prevalent technology – remote energy metering – to visualize electricity used in the household as a whole.

It is not just the light of the Flower Lamp – but its actual form – that reflects energy consumption in the home. Rather than showing how many watts are consumed at any given time, its shape is responsive to the overall trend in consumption. With a decrease in household electricity use, the Flower Lamp slowly opens up and appears to 'bloom' – small sacrifices in saving electricity or hot water are thus rewarded by a poetic change in form. If, on the other hand, energy consumption increases, the Lamp closes into a more contracted form, which also affects the quality of light emitted. Thus, both the light and form of the Lamp reflect behavioral tendencies within a household. In order to make the Flower Lamp more beautiful, a collective change in behavior is needed.

While we may not register each and every one of our personal energy-related actions – much less our habitual or household behaviors – our electricity meter does. In effect, it keeps an account of how each act of energy use builds up into more complex patterns and collective cycles of daily, weekly and seasonal energy behaviors. Remote metering allows us to access this data – or to give access to a utility company or another third party. While some of this account is available in our monthly electricity bill, presented in quantitative and economic terms to the 'head of the household', the implications may not be present in more general and ongoing family life. Five Flower Lamps are currently implemented as an add-on to current electrical meters, with the possibility for more comprehensive integration with near-future remote metering technology.

If the information presented by electricity meters and utility bills can be difficult for the whole family to relate to, the Flower Lamp gives another – inclusive and qualitative – experience of household energy use. Amplifying collective energy behaviors, the intention is to expand the presence and discussion of energy use within a household. The scale and appearance of the Lamp are related to lighting typically found in the entrance, living or dining room of a home, common spaces where families and visitors might gather on a regular basis. As a conversation piece or persistent visual reference, the appearance of the Lamp has the potential to spark and maintain interest over time.

The Flower Lamp transforms data on energy consumption into an informative and persuasive visualization in the home. Rather than simply re-producing the monthly bill in another form, the Lamp presents a more contextual and personalized value based on local behavioral change. Change is measured as a difference between the current status collected in real-time by the household electricity meter and the historical data accumulated over the long term. Taking seasonal variability into account, this results in a calculation of a net increase or decrease in energy consumption. Beyond merely visualizing the bottom-line, the Lamp amplifies overall tendencies, rewarding relative changes in energy behavior.

Even as it gives a visual form to abstract data, the Lamp is designed to be persuasive in various ways. For one thing, as a form of reward, the lamp blooms in immediate response to improvements made locally and personally, registering even small changes in behavior. However, since it rewards current behavior relative to behavior in previous months, the Lamp requires increasing improvement in order to continue to bloom. As an incentive, the threshold for blooming rises over time.

Further, the Lamp is designed as a shared representation in a common space in the home – as such, its form and behavior are up for discussion and debate among the whole family. Even as it articulates small changes and discrete choices, a dramatic change in form requires a concerted and collective effort – perhaps even some social pressure and friendly competition. Change – in the form of the Flower Lamp and in household energy behavior – must be continually and collectively negotiated.

Just as new electricity metering technology connects private households to public utilities in another way, the Lamp connects local choices to common values in a new way. Aiming to create a more nuanced expression of energy consumption in the household, the Flower Lamp is designed not only to stimulate reflection but to require action – a change in individual and collective behavior on a daily and ongoing basis. Relocating the grounds for debate on energy consumption as a family matter, made present and personal in the home, the project aims to create an everyday experience of balancing between private and public interests.

Project team: Sofia Lagerkvist, Charlotte von der Lancken, Anna Lindgren, Katja Sävström (Front Design) with Göran Nordahl

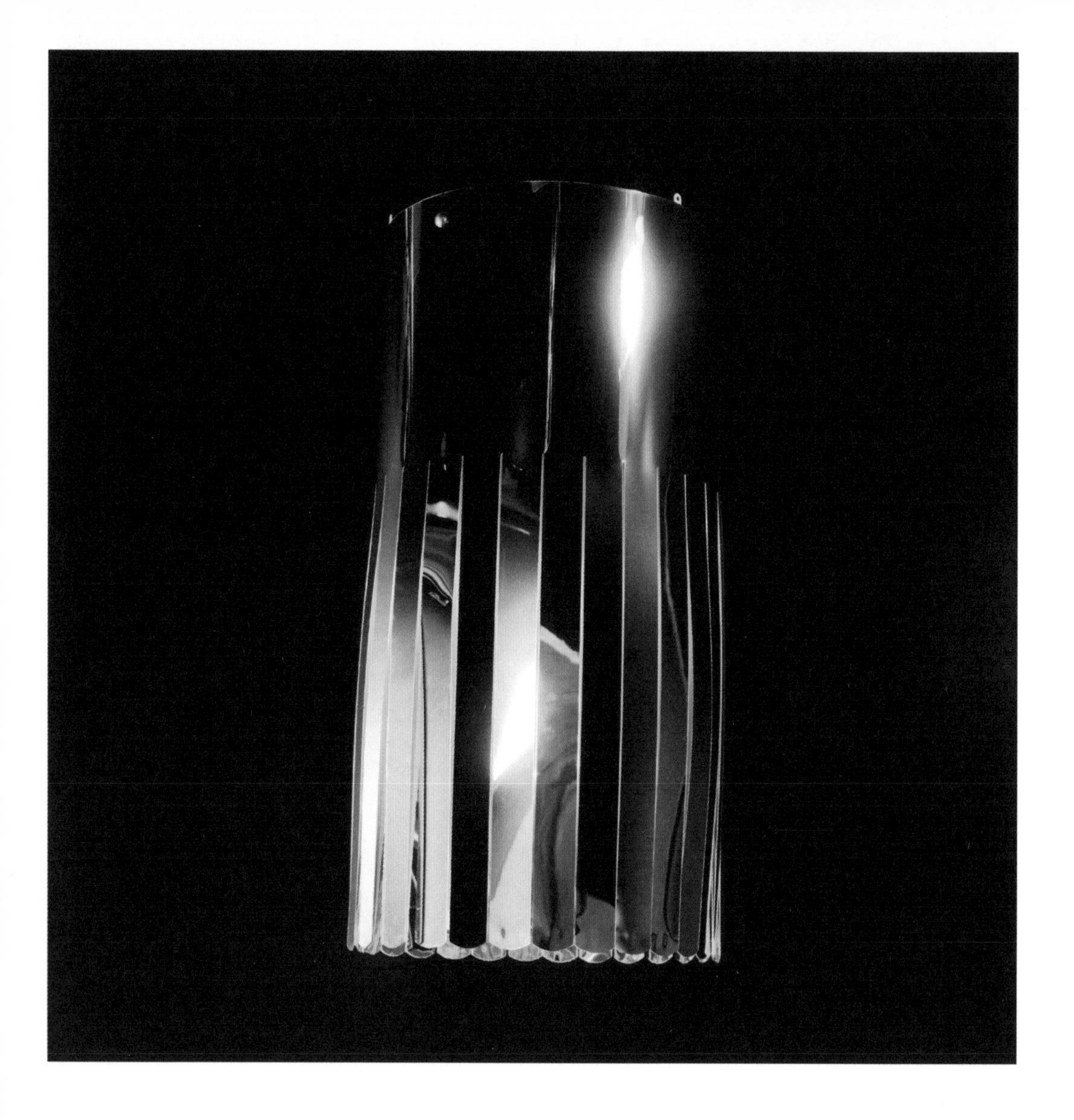

Free Energy

Many electrical appliances have become nearly indispensable to us over time. Electric kettles, irons and dishwashers transformed domestic life when they were invented – along with relations between energy producers and private consumers. Today, it is electronic products that are rapidly and radically transforming our lifestyles – and our relations to energy. As we discovered from responses to the cultural probes methods at the start of Static!, electronic products put energy into constant focus – just consider the attention paid to battery power and longevity in mobile phones! 'Free Energy' explores the extension of relationships to energy outside the home, where energy becomes not only a personal but a public issue on a daily basis.

Preliminary concepts explored potentials for em-'power'-ing ordinary situations and spaces. Speculating 'what if' solar energy collection could become ubiquitous or portable, for example, we represented potential social and spatial effects as fictional episodes in everyday life. Tea parties in neighborhood gardens? Reading lamps on public benches? Historic monuments as destinations for recharging? These speculative projections raised further questions – situating a bigger discussion about the ownership of natural resources and access to public utilities.

Rather than inventing new products, we decided to use design for opening a more public conversation about such issues. Two concepts were taken forward – not as potential products or even product proposals, but as props for inviting participation and local debate. The 'Energy Tap' and 'Kinetic Door' were two low-tech prototypes implemented and installed in public.

The Energy Tap is for making free energy. While energy typically originates in distant power plants and energy outlets are often restricted to private spaces, the Tap can be inserted anywhere so that anyone can generate their own energy. Borrowing from commercial crank radios and public utility boxes, it consists of a modular stand with a crank and an outlet. Generation requires the personal and physical effort of cranking, and ownership over the resulting energy is local and direct. Energy production and consumption become a public spectacle, along with new activities and actors that might emerge.

Marlb
LÖRDAG
Nyhet
Rea
20%
GT
PREMIÄR
i dag för GT:s nya
MOTOR-
TIDNING
Du får den på köpet!
AFTONBLADET
TJEJER
i nya tv-såpan
HAR
ANOREXI
TV
SNABBARE
DATOR
PÅ 20
MINUTER
8 enkla steg
STADSJEEPAR

The entrance doors to buildings impact heat and, thus, energy conservation. In many buildings, sets of ordinary, automatic and revolving doors may be placed next to one another, symbolizing a range of more (or less) eco-friendly choices. Intervening into this choice, the Kinetic Door attaches onto any revolving door to reward door-pushers for making an extra effort. It consists of a small wheel and LED lights – the wheel rotates when the door revolves, powering the lights via a hacked dynamo. Thus, pushing the door sparks an aesthetic lighting pattern, siphoning personal effort into a visible reward.

These were a basis for staging interventions in four different local spaces. As props, rather than products, the prototypes prompted both a physical and conceptual exploration of energy behavior. In conversations sparked with one another and in follow-up interviews with us, people's responses spanned from the utilitarian to the utopian. About the Tap, one noted, "I would use it to recharge my car or my future car," while another proposed, "I love the idea of free energy that is all about releasing energy into the world... with free energy, people would be more connected... maybe people would get out of their houses and throw parties in the street and get to know each other better." Undercurrents of ethics and agency also emerged – while the idea of energy for free prompted some to imagine applications for those in need, others saw the chance to create surplus energy for care-'free' activities.

Drawing on strategies from art and activism, Free Energy draws attention to the conventions designed into everyday life. The Kinetic Door, for example, is a 'parasite' on the official architectural boundary between public and private space – a site where personal choices directly affect environmental conditions. By highlighting or adding choices to the status quo, the intention is both to stimulate reflection and debate, and also to give a tangible demonstration of alternatives. Amplifying personal effort in the public realm ¬– the outcome is not only an individual benefit but, potentially, a social effect. Since the prototypes embody other ways of thinking about access and control, effort and effect, use involves people directly and publicly in the production of another 'balance of power'.

Project team: Anders Ernevi, Margot Jacobs, Ulrika Löfgren with Sara Danielsson (School of Design and Crafts at Göteborg University)

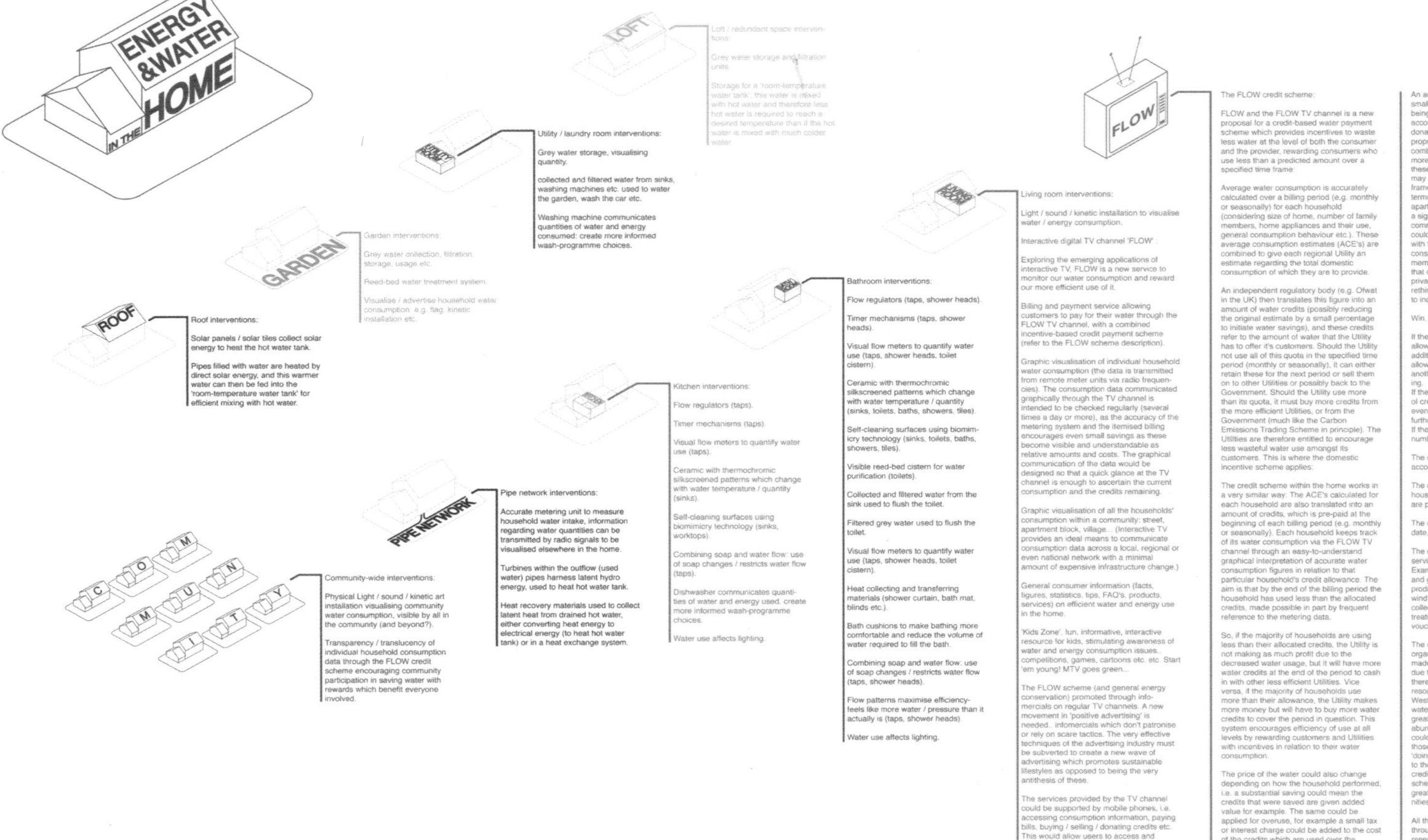

The FLOW credit scheme:

FLOW and the FLOW TV channel is a new proposal for a credit-based water payment scheme which provides incentives to waste less water at the level of both the consumer and the provider, rewarding consumers who use less than a predicted amount over a specified time frame:

Average water consumption is accurately calculated over a billing period (e.g. monthly or seasonally) for each household (considering size of home, number of family members, home appliances and their use, general consumption behaviour etc.). These average consumption estimates (ACE's) are combined to give each regional Utility an estimate regarding the total domestic consumption of which they are to provide.

An independent regulatory body (e.g. Ofwat in the UK) then translates this figure into an amount of water credits (possibly reducing the original estimate by a small percentage to initiate water savings), and these credits refer to the amount of water that the Utility has to offer it's customers. Should the Utility not use all of this quota in the specified time period (monthly or seasonally), it can either retain these for the next period or sell them on to other Utilities or possibly back to the Government. Should the Utility use more than its quota, it must buy more credits from the more efficient Utilities, or from the Government (much like the Carbon Emissions Trading Scheme in principle). The Utilities are therefore entitled to encourage less wasteful water use amongst its customers. This is where the domestic incentive scheme applies:

The credit scheme within the home works in a very similar way: The ACE's calculated for each household are also translated into an amount of credits, which is pre-paid at the beginning of each billing period (e.g. monthly or seasonally). Each household keeps track of its water consumption via the FLOW TV channel through an easy-to-understand graphical interpretation of accurate water consumption figures in relation to that particular household's credit allowance. The aim is that by the end of the billing period the household has used less than the allocated credits, made possible in part by frequent reference to the metering data.

So, if the majority of households are using less than their allocated credits, the Utility is not making as much profit due to the decreased water usage, but it will have more water credits at the end of the period to cash in with other less efficient Utilities. Vice versa, if the majority of households use more than their allowance, the Utility makes more money but will have to buy more water credits to cover the period in question. This system encourages efficiency of use at all levels by rewarding customers and Utilities with incentives in relation to their water consumption.

The price of the water could also change depending on how the household performed, i.e. a substantial saving could mean the credits that were saved are given added value for example. The same could be applied for overuse, for example a small tax or interest charge could be added to the cost of the credits which are used over the original estimate. These measures could add further incentive to waste less around the home.

An additional aspect of the scheme is for a small percentage of saved household credits being automatically donated to a community account. Subsequent credits can also be donated to the community account. This proposal encourages co-operation, as combining credits as a community gives much more buying power when trading or exchanging these for products and/or services. A family may only make a small saving over the time frame and this may not amount to much in terms of exchange or donation, but a whole apartment block of similar savings could create a significant total to be exchanged as a community for the benefit of all involved. This could also operate through the TV channel, and with the transparent / translucent community consumption information, every community member's water use is on show for everyone in that community to see. This gentle invasion of privacy may be enough to encourage people to rethink their consumption patterns, not wishing to incur the wrath of their gossiping neighbours!

Win, Lose or Tie:

If the household has used more than it's credit allowance, it must pay for the excess use in addition to the next billing period's credit allowance cost, or buy / trade credits from another household which has credits remaining.
If the household uses exactly the same amount of credits as it was allocated, it has broken even: there is no remaining credit, and no further charges to pay.
If the household has credits remaining, it has a number of options:

The credits can be donated to the community account.

The credits can be traded with / sold to other households / businesses / organisations that are part of the FLOW scheme.

The credits can be saved and used at a later date, to cover overuse for example.

The credits can be exchanged for products and services which promote sustainable lifestyles. Examples could be: organic food, books, toys and games, more efficient / greener household products, communal-use tools and appliances, wind turbines, solar panels, grey water collection and reuse systems, reed bed water treatment systems, sustainable transport vouchers, etc.

The credits can be donated to a charity / aid organisation. As the monetary savings to be made from using less water are relatively small due to the low price of water, the scheme therefore needs to add further value to this resource and the need for conservation. As The West is water rich, relatively small amounts of water mean very little to us but could mean a great deal more to those with much less abundant supplies. Savings made in the home could be donated in the form of water credits to those who need them more, and the feeling of 'doing the right thing' becomes more valuable to the consumer than the actual value of the credits he/she is donating. An entire trading scheme of water credits could exist enabling greater efficiency and fairness across communities, countries, continents...

All these choices are made and carried out through the FLOW TV channel. Users can research each option and make informed choices regarding their credits.

Energy systems do not operate in isolation. Considering that up to 20% of household electricity is used to make hot water, it becomes obvious that electrical and plumbing systems are interdependent from the perspective of energy use. As we collected our thoughts at the end of Static!, our collection of design examples seemed to highlight the inevitable interdependency of different systems. Systems of electricity and water, electric appliances and electronic devices – even patterns of personal behavior and family activities – all impact upon one another. Just as 'no man is an island' in a sustainable world view, every designed thing belongs to an larger and evolving ecology of things, at home and beyond.

Flow

Expanding on this, we developed 'Flow' to give a more extensive sense of the use – and implications – of one such energy system. Flow is a service that gives a report and choices about how water is used in the home. Just as remote electricity metering allows utility companies to monitor energy use at a distance, Flow is based on remote water metering technology. Thus, measurements of water consumption would be transmitted wirelessly, fed back into an enabled television – or even to a computer or mobile phone. Integrated with digital and interactive television, this data would be visualized and constantly updated to provide households with a real-time view of their water use.

Subscribers to the Flow service would be allocated a certain amount of water each month, a calculation based on household size, location, season and other related factors. This standard amount is pre-paid at the start of each month. Each day, the family can tune into a TV station where their water consumption is portrayed as a visual meter. Clearly delineated in columns of 'used' and 'unused', the interface shows their actual water consumption compared to their pre-paid allocation. Thus, it indicates the status of the household allowance day-by-day over the course of the month. Additionally, personalized tips and other forms of information and entertainment about energy are available at a click of the remote control.

If a family consumes more than usual, their TV interface and next billing cycle reflects the increase in use – and in cost. However, if consumption decreases, the water that is conserved can be put to use in several ways. It can be saved, for example, in a special account to compensate for any future over-consumption. Or, the value of water saved can be exchanged for sustainable products offered by service partners – thus perpetuating a cycle of 'green consumerism'. Besides recycling savings into further consumption, the family can also decide to donate the value of their water conservation. Donated to community groups or non-governmental organizations (NGOs), household effort can be extended further to support sustainable initiatives locally or in other parts of the world. Thus, Flow encourages local change by an incentive-based pricing scheme and philanthropic motives – even neighborly competition!

YES

Flow is designed to increase awareness, direct choice and economic rewards with respect to household energy use – thereby serving as a reminder not only of the economic cost but the human and environmental value of decreasing domestic energy consumption. Indeed, different forms of current (electric and water) are literally earned, exchanged and spent as currency. Weaving together diverse motives, stakeholders and forms of currency, Flow is an attempt to balance the often competing interests of social, environmental and economic sustainability. In reality, of course, this would present a challenge to develop a new kind of business model for combining technologies and partners. However, it is precisely this integration that makes such a 'service ecology' meaningful – monetary and societal value become interdependent both on a household and a humanitarian level. Participating individuals and families join into a system of other actors, encouraging the tendencies of other families, communities and organizations in lean thinking, natural capitalism and corporate responsibility.

As a culmination of many of our findings over the course of Static!, Flow was developed through discussions that served both to conclude the research program and to suggest potentials for future research. The service was developed as diagrams, drawings and slideshows to explain the connection of various technical and energy systems. In addition, a short movie was made in conjunction with a local family to explore a potential scenario of use. Various emotional and behavioral responses were explored through this collaboration, and were explained in a narrative form that could be used to document and communicate the project to a wider audience. Flow was further developed at the Interactive Institute in a spin-off project called 'Wattch!' and in the next generation of Static! called 'Aware', which framed its initial investigation in terms of the systems perspective.

Project team: Alex Allen, Anders Ernevi, Margot Jacobs, Ulrika Löfgren

FLOW
more for using less

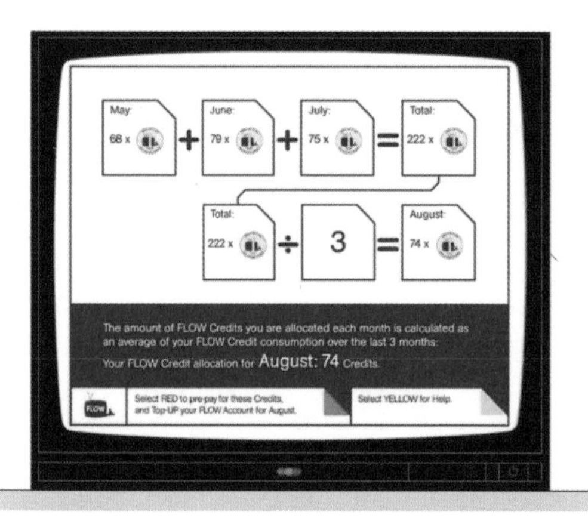
May: 68 x
June: 79 x
July: 75 x
Total: 222 x
Total: 222 x
3
August: 74 x
The amount of FLOW Credits you are allocated each month is calculated as an average of your FLOW Credit consumption over the last 3 months:
Your FLOW Credit allocation for August: 74 Credits.
FLOW
Select RED to pre-pay for these Credits, and Top-UP your FLOW Account for August.
Select YELLOW for Help.

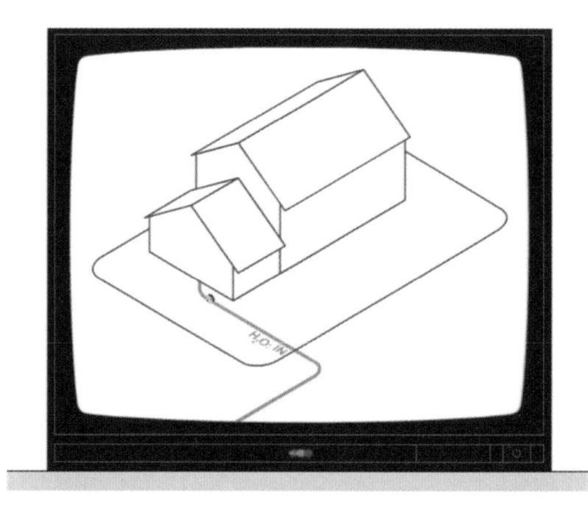
H₂O: IN

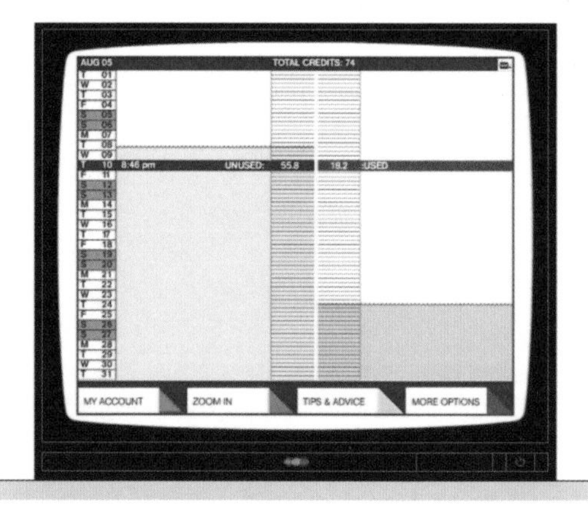
AUG 05
TOTAL CREDITS: 74
T 10 8:46 pm
UNUSED: 55.8
18.2 :USED
MY ACCOUNT
ZOOM IN
TIPS & ADVICE
MORE OPTIONS

Perspective

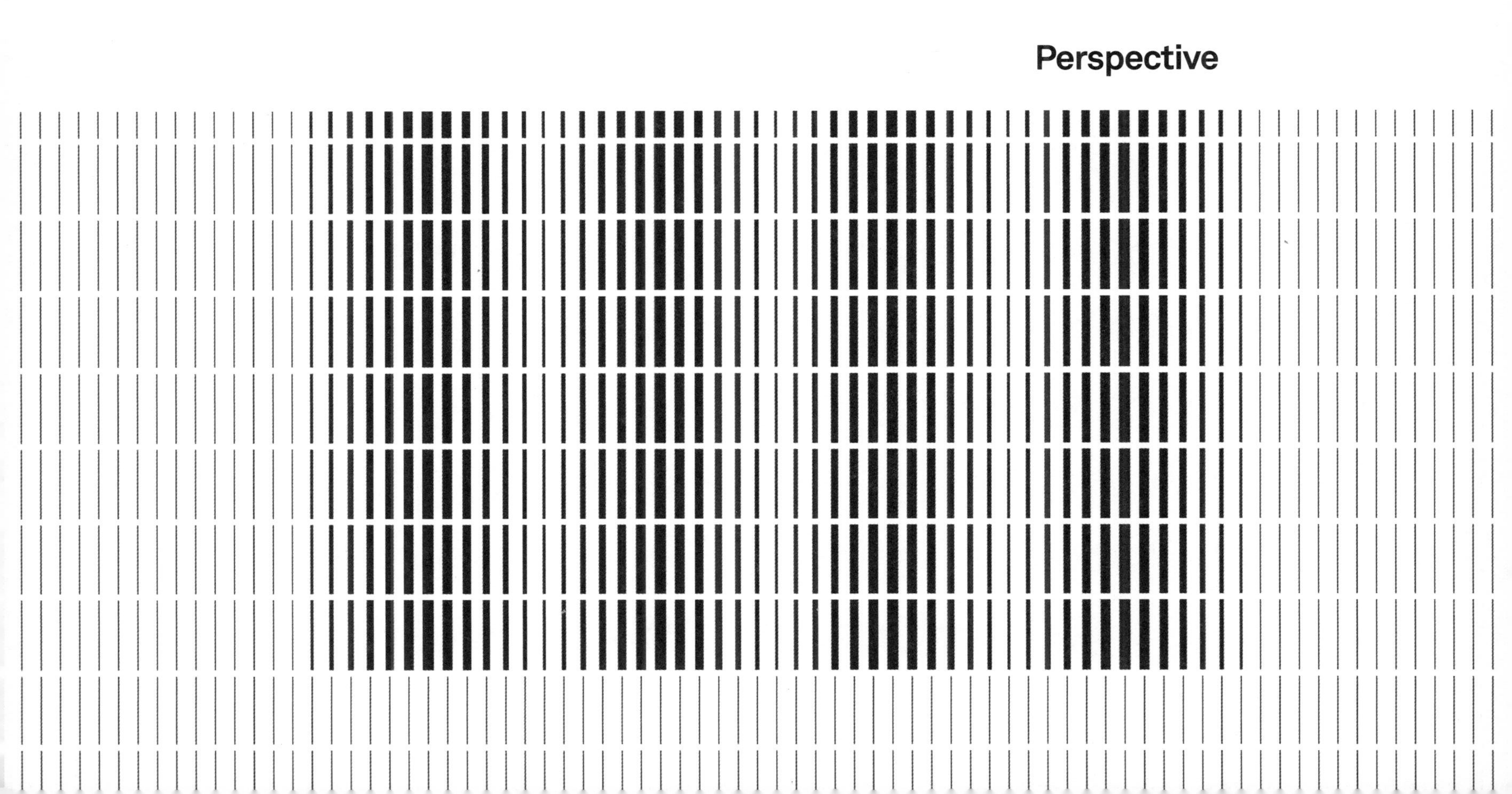

Static! at home
Sara Routarinne

"The built environment... is the physical infrastructure that enables behavior, activity, routines, habits and rituals".[1] Within such infrastructures, 'home' is a special kind. Not only a man-made shelter or an architectural construction, it is a meaningful place that emerges in the activity of living. The emergence and maintenance of a home involves its inhabitants, their social relations, and their practices – the ways in which they use and organize physical objects and the environment to lead their lives. Even if the lifeworld is furnished with objects designed to enable these projects, designers remain incapable of designing the everyday.

Little do we know about the forces that take effect when a newcomer arrives to a household. When a baby is born, the family dynamics change. Likewise, when an interactive – or static – artifact is adopted in a home, the configuration of domestic practices is put into flux. The newcomer must compete, or lose its position, in the configuration of a home. In this process, it is difficult to predict if and how design intentions are established, mobilized or manifest in actual use. In any case, use over time may alter first impressions, for better or worse. Therefore, an extended period of use in a real-life context provides an intriguing field to experiment with products.

As an appendix to Static!, two of the early prototypes were offered for households to experience. The goal was to investigate whether the prototypes, which were designed to make an argument about energy awareness, succeeded in making it. We call this endeavor 'domestication as design intervention', in which a conceptual design is intervened into the pre-existing practices of a household.[2]

Domestication as design intervention

Within the social sciences, 'domestication' refers to processes of differentiation in forms of consumption within individual households. In the context of a growing research interest in consumption and everyday life, the domestication approach was established by Roger Silverstone and his colleagues during the 1980s to address how households with similar socio-economic backgrounds buy and enjoy different things.[3] Researchers entered through the closed doors of private homes to try and understand the processes involved in 'taming' artifacts, or how users engage mentally and physically in actively constructing the meanings of things. As an approach to qualitative research, domestication is sensitive to generational, gender, moral and other issues, since these influence the manifestation of values in consumption choices and practices.[4]

During a process of domestication, a new artifact is understood to challenge and, over time, find its slot within a particular material and social 'ecology'. In the field of design inquiry, the domestication approach has been applied to study the influences of design in and on use. Understood as interventions into existing situations and contexts, different types of design materials and objects can be introduced to open up design ideas or to pose research questions to users.[5]

For example, Forlizzi and DiSalvo conducted a study of the Roomba vacuum, in which a wireless floor-cleaning robot was domesticated within fourteen households.[6] They investigated how the technology met expectations in a domestic context, how it became socialized, and how homes and domestic appliances adapted to one another. With respect to expectations, the intelligence of the vacuum was a disappointment, though users were surprised by the quality of cleaning. As to socialization, Roomba changed the division of labor in the households – males took a lead in floor cleaning, reversing an existing gender bias. As to adaptation, the households had to reorganize, whether this meant removing objects that were obstacles in the Roomba's way or creating obstacles to block off places where it might fall or bump into precious objects.

Such findings prompted the researchers to articulate the design challenges that became recognizable: the challenge of introducing new technologies for consumers through social, rather than technological, arguments, the challenge of impacting practices within a household, and the challenge of understanding the home as an ecology of material and social entities.

Taking Static! home
A similar line of enquiry was pursued with respect to two prototypes – the 'Erratic Radio' and the 'Energy Curtain' – from Static! After the program, these temporarily immigrated to Finland to be a part of another research study that I conducted based on the domestication approach.

One difference from artifacts in related studies is that the Radio and the Curtain are less purposeful and more conceptual than the Roomba. They are perhaps more in line with the experiments of Gaver et al.[7] and the conceptual products introduced in *Design Noir.*[8] Since the artifacts they created could not be understood only in terms of purpose or function, they steered users into domestication by requiring interpretation. In fact, it was the unexpected functional – or dysfunctional – traits that put the people's minds to work as they tried to make sense of the prototype (Gaver et al.). These accounts inspired our research design with the Erratic Radio and the Energy Curtain.

Conducted as a set of field experiments, the study deployed both prototypes over a period of three to five weeks each in four different Finnish households. Information was gathered with pre- and post-interviews and diaries. A trigger sheet was used in the pre-interview when a prototype was introduced to a family *(see the image on page 105)*, in which they were asked to locate themselves on axes of sustainability and technology acquisition. During the domestication period, they kept a diary and/or e-mailed comments and questions. At the end, the households were interviewed on their experiences with the prototype in question and the effect upon their energy awareness.

Ympyröi testattava STATIC! -prototyyppi:

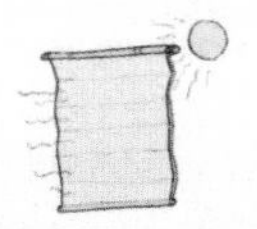

Erratic Radio - Oikutteleva radio **Energy Curtain** - Energiaverho

Koosta taloutenne henkilömäärä kuvia ympyröimällä:

Ympyröi ikäsi ja sukupuolesi:

alle 20 20-27 28-35 37-43 44-51 52-59 60-67 68-75 yli 76 **M N**

Miten sijoittaisit itsesi oheiseen taulukkoon, jossa vaaka-akselilla arvioit omaa suhtaumistasi ekologisiin kysymyksiin ja pystyakselilla suhdettasi uuteen teknologiaan?

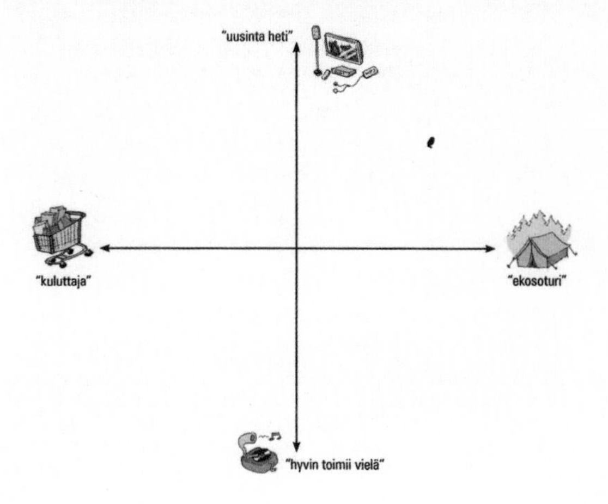

Findings from the study
A variety of early impressions proved to affect household perceptions. It is critical how a new technology is introduced (see Forlizzi and DiSalvo), which was obvious with the Erratic Radio. The concept of 'erratic' does not directly translate into Finnish, and can carry negative connotations. While the Curtain was warmly welcomed, until the translation *oikukas* ('capricious', 'moody') was found, nobody agreed to test the Radio. Initial use also exposed issues. The Curtain was able to make its point – although not exactly as designed. One domesticator writes about her first night: "I turned off the light, I even closed the door and sat on the bed experiencing that the curtain would glow. I watched and watched and was imagining seeing something but it was probably only an illusion. The curtain was dark, and I was pretty disappointed." There was too little daylight in the Finnish winter to support this function, much less immediate user expectations.

In fact, the Curtain made another respondent realize how dark the winter is – darkness became translated into lack of energy. This reflection, however, did not seem to aid sustainability, since she found an energy-consuming solution to supplement the Curtain *(image on page 108)*. Yet, when daylight increased in the spring, the Energy Curtain operated in symbiosis with a computer in one home, providing welcome shade for the monitor. The energy cycle of the Curtain was restored, and it glowed as intended. However, as summer approached, there was no use for the its light during the bright nights. By drawing people's attention to such seasonal relations to light, and personal dependence upon this, the Energy Curtain did indeed raise energy awareness in the long term.

As for the Erratic Radio, it was more of a challenge to understand. All domesticators complained that it did not make itself clear, that they were not able to see how its unreliable functionality might comment on their use of electricity. Two types of responses were found. On one hand, the Radio was not able to keep the interest of the adult members in two different households. They abandoned it after a few days, and it was eventually adopted by the youngest members of each family. On the other hand, the Radio was encountered as a challenge in two other households consisting of couples without children – each set out to detect the mystery behind the erratic functionality. Beyond design intentions, this seemed to indicate the role of different and evolving intentions in use.

Further, the Radio made all the domesticators very aware of electric and electromagnetic radiation. One of the couples wondered whether the Radio would stay tuned in the bathroom, and the other couple took it to their bathroom *(image on page 107)*, even wrapping it in aluminum foil to test whether this Faraday cage would make it more reliable. Such experiments raised further questions, since influences on the radio traveled through walls, doors and windows. One wondered about relations to her neighbors through the Erratic Radio: "I don't know what are the appliances it responds to, but I mean, if we have nothing electrical on and it still starts to buzz then is it the appliances of the neighbors? Like, should we go and ring the doorbell and tell them to use less electricity?"

°C
50°
40°
30°
Min
1000
900
800
500
Whirlpool
40°

Reflections

Previous inquiries clearly indicate that introducing semi-functional, unfamiliar objects into a familiar everyday context, and leaving them there for a while, is an effective way to provoke (see Dunne and Raby). An unidentified object helps people to reflect upon their experiences, desires and values. For designers, such information is a source of inspiration.

Another important reason for exploring and developing this kind of study in design is that conceptual products more profoundly illustrate the issue of how new things do not exist in a void but, rather, are brought into established systems of objects already appropriated. This perspective is too easily lost in more traditional usability evaluations. Such evaluations are usually conducted as controlled experiments with the aim of testing functionality. However, especially with the early prototypes from Static!, the design objectives did not concern functionality but the exploration of people's energy awareness and, potentially, even change in their energy behavior. Therefore, traditional tests would not have been very useful.

Based on our field study, form and form-giving of the Static! prototypes had influences on people's energy awareness. In a domestication experiment, design prototypes can act as arguments, triggers or teasers for users to become more aware of their dispositions. The diverse and evolving relations to the Static! prototypes nicely demonstrate the complexity of domestication processes, variable across contexts and seasons, households and family members, and over time in use.

Notes

1. Hunt, "Just Re:Do it," in *Strangely Familiar*, ed. Blauvelt, 58.
2. See Routarinne and Redström, "Domestication as Design Intervention," in *Proceedings of NORDES*.
3. For further background, see Silverstone, Hirsch and Morley, "Information and Communication Technologies," in *Consuming Technologies*, ed. Silverstone and Hirsch; Silverstone, *Television and Everyday Life*.
4. For further background, see Silverstone, Hirsch, and Morley, "Information and Communication Technologies"; Berker, Hartmann, Punie and Ward, "Introduction," in *Domestication of Media and Technology*.
5. For a more thorough discussion, see Routarinne and Redström, "Domestication."
6. See Forlizzi and DiSalvo, "Service Robots in the Domestic Environment," in *Proceedings of HRI*.
7. See Gaver, Bowers, Boucher, Law, Pennington and Villar, "The History Tablecloth," in *Proceedings of DIS*.
8. See Dunne and Raby, *Design Noir*.

Tatu Piispanen created the worksheet and assisted in the interviews.

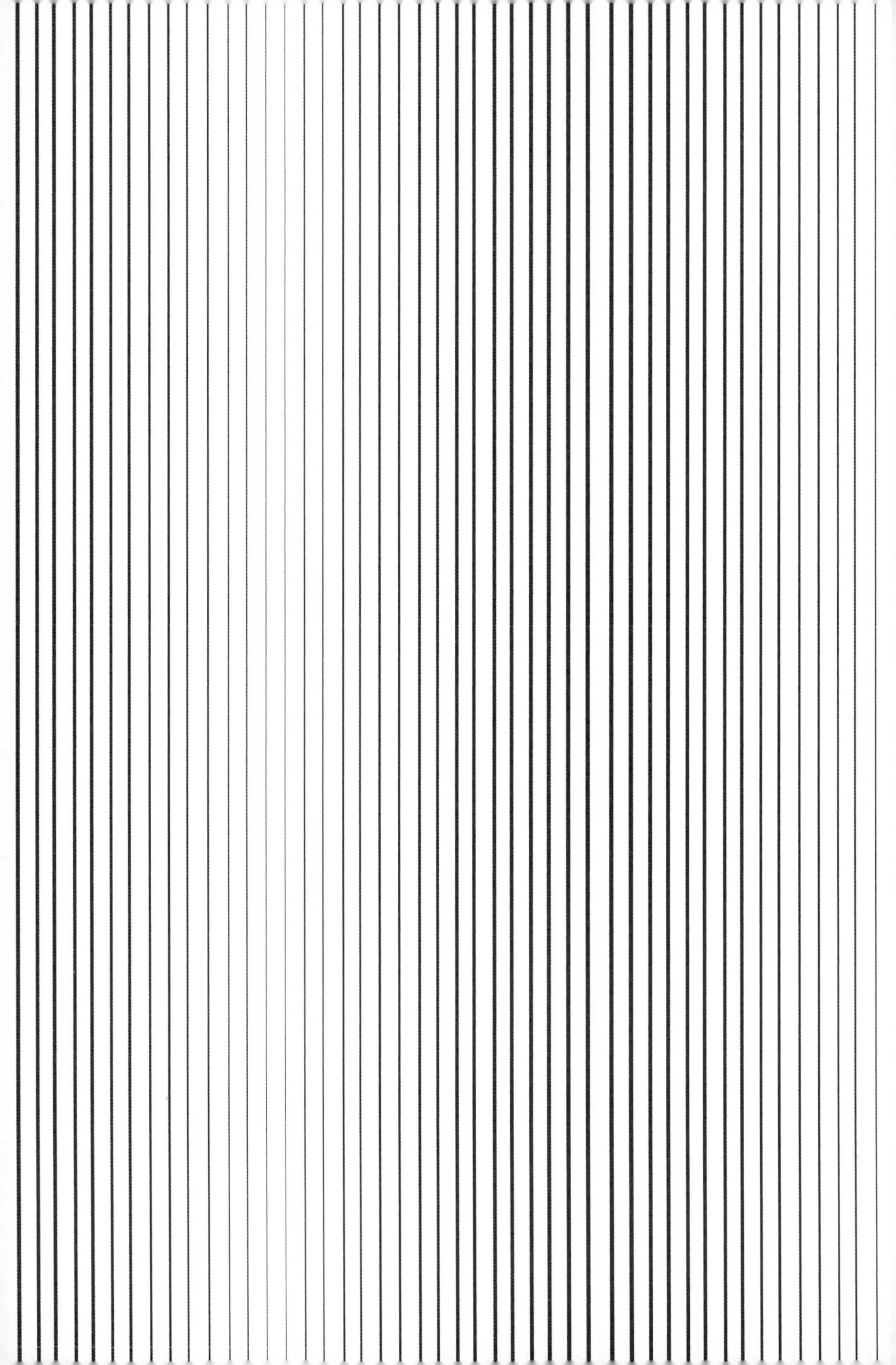

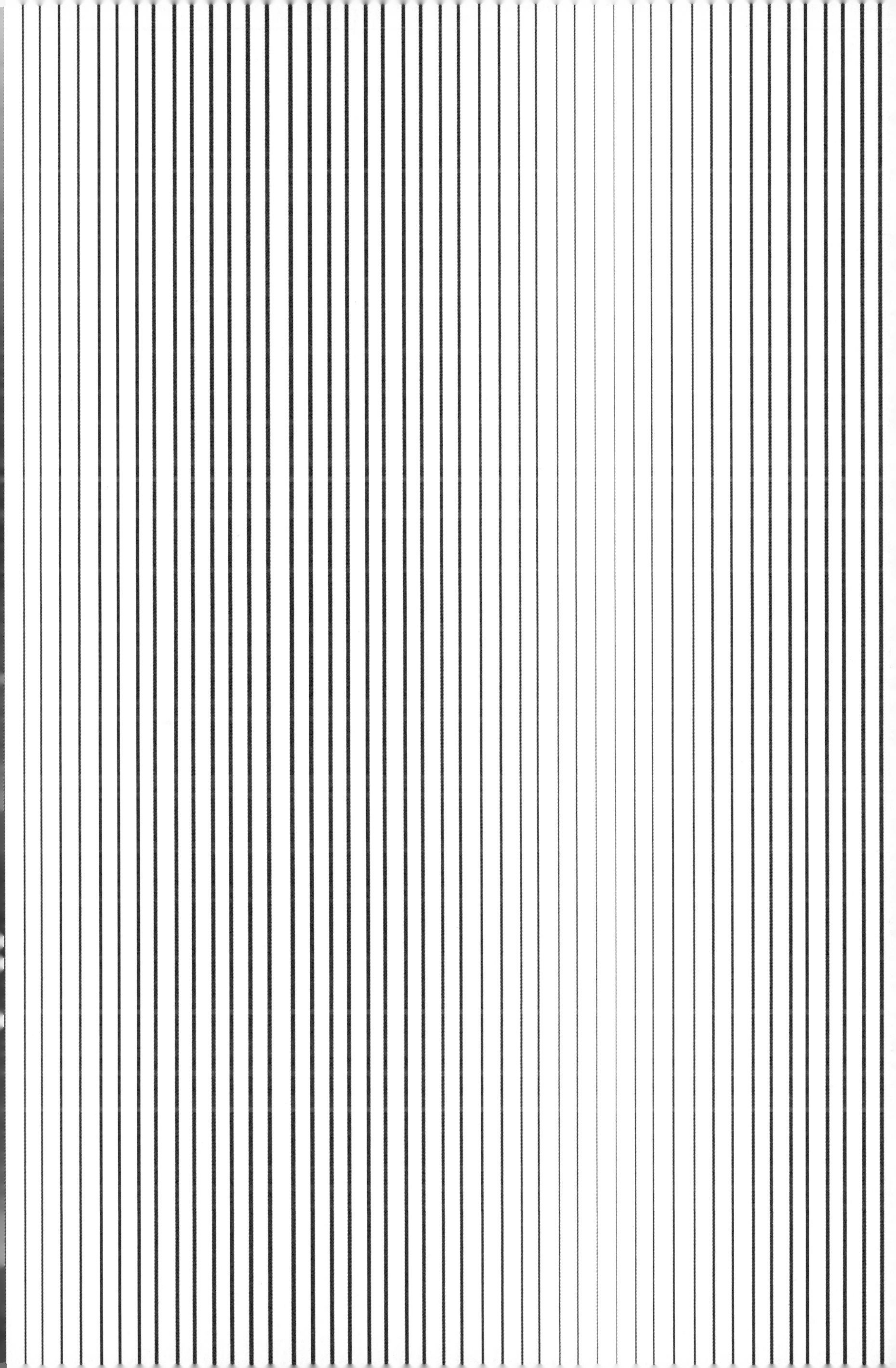

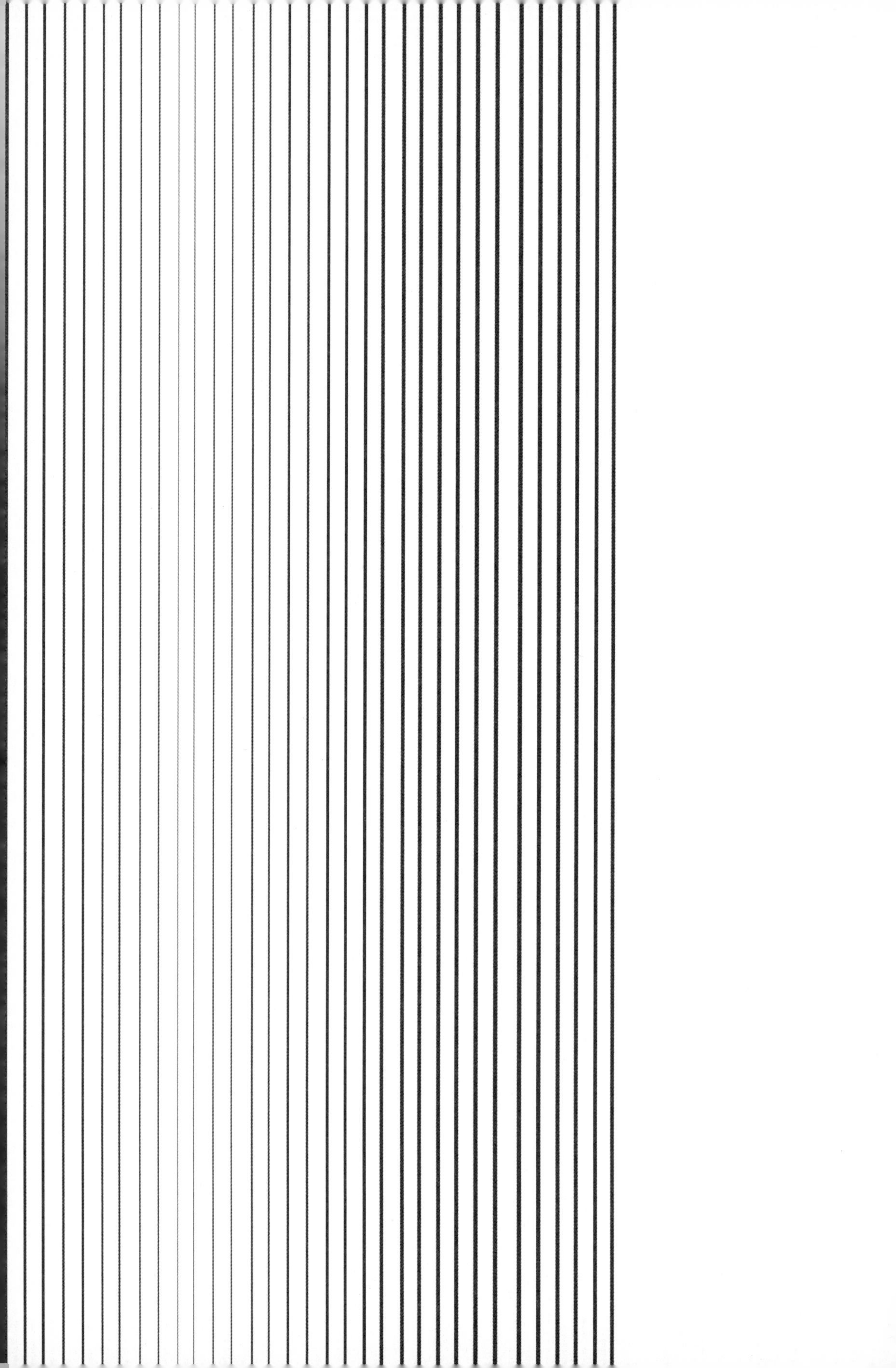

Concluding remarks

During the project and since its completion in 2006, Static! has been successful in fostering reflection and debate about energy awareness – among designers, users, and a range of other stakeholders and audiences. Evident in the evaluation studies, the design examples have been proven to prompt changes in perceptions and behaviors with respect to energy use. As open-ends for investigating relations between design and energy awareness with our collaborators, the processes and products of Static! have had an impact on continuing work at our academic and industrial partners, including current doctoral projects and commercialization initiatives. Widely exposed internationally, the design examples and the meta-design of the research program have been a platform for hosting a wider debate about the values, ideologies and implications of design and technology on sustainability and consumption.

There has been enormous interest among the public and consumer groups. Repeated appearances at Energitinget, the largest energy event in Northern Europe, and at *Wired Magazine*'s NextFest expo have opened a communication channel to the top international magazines, TV shows, and radio programs, including a mention as one of *TIME Magazine*'s Best Inventions from 2006. This coverage has provided insights into how the public perceive and relate to sustainability and to design research as well as sparking valuable collaborations. A second component has been two extended studies of some design examples carried out within households, one in collaboration with the University of Art and Design in Helsinki, Finland, and another through the Department of Technology and Social Change at Linköping University in Sweden.

The findings – and questions – generated in Static! continue to influence our work. Indeed, Static! was extended 'virtually' through the long-term domestication studies, the results of which are still in the process of being disseminated. The effect has been a continuing stream of ideas and inspirations interjected into our work in other projects – as well as prompts for new follow-up projects. Static! has been followed by several projects within the Interactive Institute investigating similar questions. 'Aware', launched in 2006, was motivated by a need to locate design within a more systemic understanding of energy and behaviors in households. Initiated in 2008, 'Switch!' expands beyond discrete products and households to consider architectural and urban interactions in which community and social design become important factors.

This impact of Static! has helped sustain – and grow! – commitment to the research area. The attention has helped us to raise awareness about and methods for conducting a critical and design-oriented inquiry into energy and sustainability issues. In addition to a nuanced and first-hand view of design and design research, the Energy Agency has also been encouraged by public, consumer, and commercial interest. Adding to their core set of research programs, the agency has launched the area of 'Design, Energy and IT', which has funded further research at the Interactive Institute as well as stimulating this interest among other institutes and universities by extending the scope of funding. Knowing that Static! has inspired and influenced so many represents perhaps our most significant result – and contribution to what we hope is a more sustainable future.

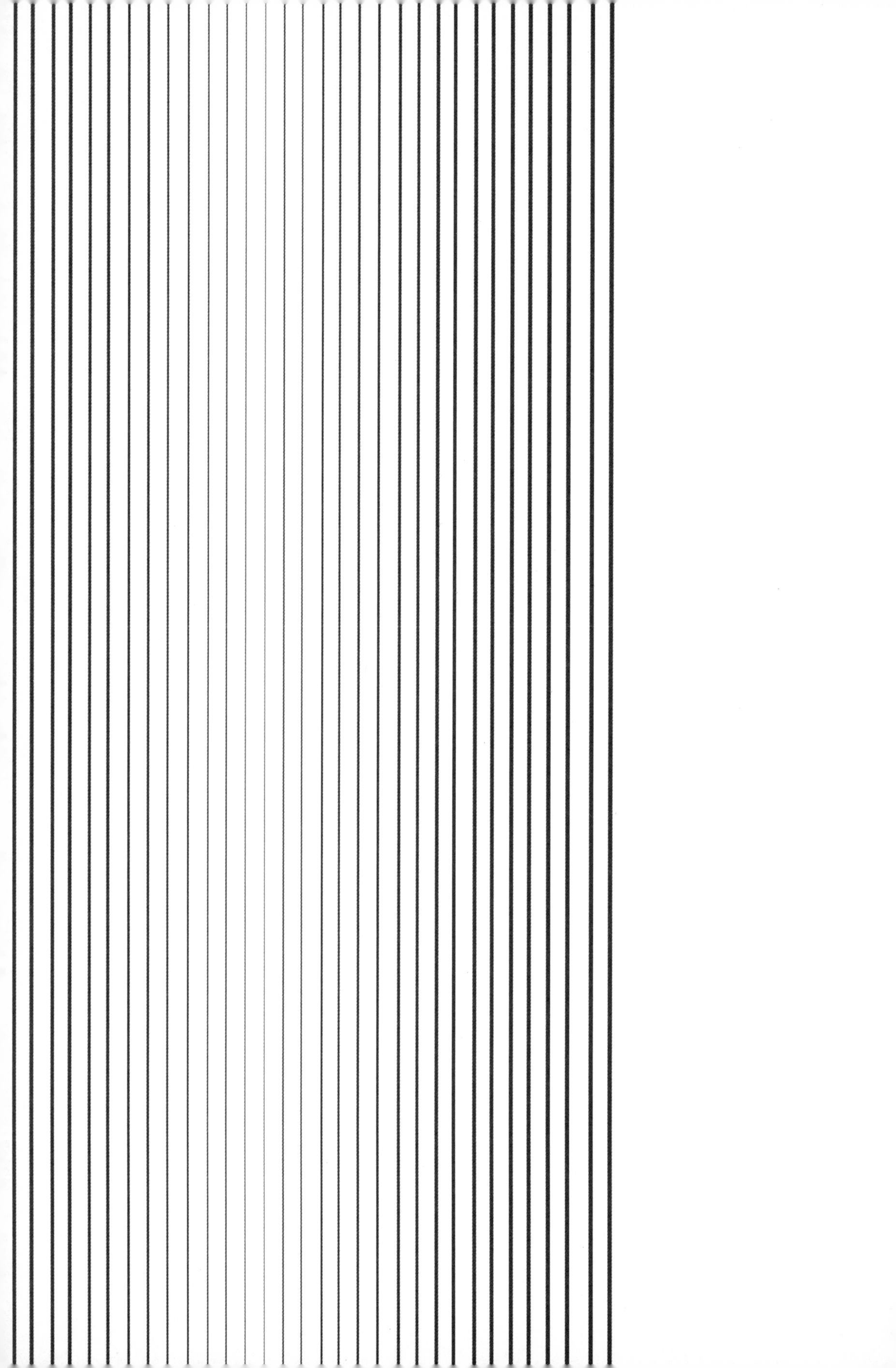

Project participants

Funding

Static!, carried out between 2004 and 2006, was funded by the Swedish Energy Agency (Energimyndigheten), with additional support from Region Västra Götaland through the Energy+Design Network.

Project lead

Christina Öhman

Research directors

Sara Ilstedt Hjelm / Ramia Mazé / Johan Redström

Project team

Alex Allen / Sara Backlund / Anders Ernevi / Anton Gustafsson / Magnus Gyllenswärd / Sara Ilstedt Hjelm / Margot Jacobs / Caroline Karlsson / Ulrika Löfgren / Ramia Mazé / Carolin Müller / Samuel Palm / Johan Redström / Christina Öhman

Doctoral students

Aarhus School of Architecture, Denmark Andreas Lykke-Olesen / *Department of Public Technology, Mälardalen University* Andreas Kvarnström

Master students

School of Design and Crafts at Göteborg University Sara Danielsson, Kerstin Sylwan / *IT University Göteborg* Mattias Ludvigsson / *Swedish School of Textiles at the University College of Borås* Carolin Müller, Linnéa Nilsson / *Royal Institute of Technology (KTH)* Kari Leppimaa

Project partner

Front Design Sofia Lagerkvist, Charlotte von der Lancken, Anna Lindgren, Katja Sävström

Project collaborators

Danish Center for Design Research, Copenhagen / *Swedish Energy Agency* Stefan Jakélius / *Göteborg Energi* / *Imego Institute of Micro and Nanotechnology* / *IT University Göteborg* / *JLT Lighting* / *Linköping University, Department of Computer and Information Science* Magnus Bång / *Linköping University, Department of Technology and Social Change* Erica Löfström / *Ludvig Svensson,* Olle Holmudd, Pepe Comi / *Mälardalen University* Anna-Christina Blomkvist, Erik Dahlquist, Thomas Karlsson, Fredrik Wallin / *School of Design and Crafts at Göteborg University* / *Swedish Industrial Design Foundation (SVID)* / *Swedish School of Textiles at the University College of Borås* Linda Worbin / *Thinlight* / *University of Art and Design Helsinki, Finland* Sara Routarinne

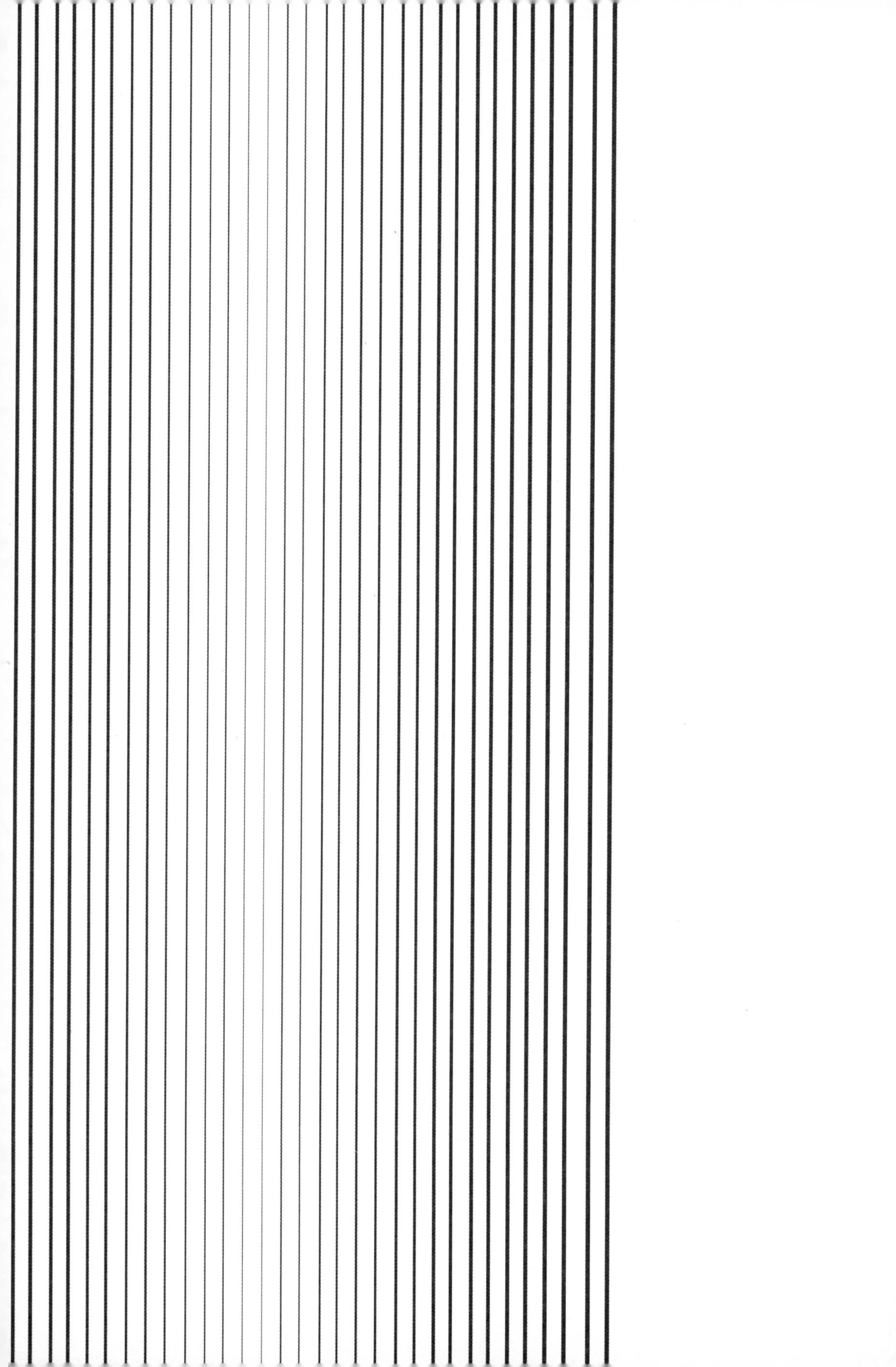

Bibliographic references

General references on Static!

Backlund, Sara, Anton Gustafsson, Magnus Gyllenswärd, Sara Ilstedt Hjelm, Ramia Mazé and Johan Redström.
"Static! The Aesthetics of Energy in Everyday Things."
In *Proceedings of the Design Research Society conference (Wonderground)*. Lisbon, Portugal: IADE, 2006.

Broms, Loove, Magnus Bång and Sara Ilstedt Hjelm.
"Persuasive Engagement: Exploiting lifestyle as a driving force to promote energy-aware use patterns and behaviours."
In *Proceedings of the Design Research Society Conference*. Sheffield, UK: DRS, 2008.

Gustafsson, Anton.
Positive Persuasion: Designing enjoyable energy feedback experiences in the home.
PhD dissertation in Applied Information Technology, Göteborg University, 2010.

Ilstedt Hjelm, Sara.
"Energi som Sens."
In *Under Ytan: En Antologi om Designforskning*, edited by Sara Ilstedt Hjelm, 118-131. Stockholm: Raster, 2007.

March, Wendy, Margot Jacobs and Tony Salvador.
"Designing Technology for Community Appropriation."
In *Proceedings of the conference on Human Factors in Computing Systems (CHI),* 2126-7. New York: ACM Press, 2005.

Mazé, Ramia.
"Criticality Meets Sustainability: Constructing critical practices in design research for sustainability."
In *Proceedings of Changing the Change*. Turin, Italy: Allemandi Conference Press, 2008.

Mazé, Ramia.
Occupying Time: Design, Technology and the Form of Interaction.
PhD dissertation in Interaction Design, Malmö University / Blekinge Institute of Technology. Stockholm: Axl Books, 2007.

Mazé, Ramia.
"Joining Tradition and Innovation in Design Research Projects."
In *Proceedings of the conference 'Joining Forces' at the World Design Congress*. Helsinki: ERA, 2005.

Mazé, Ramia and Johan Redström.
"Difficult Forms: Critical practices of design and research."
Research Design Journal 1, no. 1 (2009): 28-39.

Mazé, Ramia, and Johan Redström.
"Difficult Forms: Critical Practices in Design and Research."
In *Proceedings of the conference of the International Association of Societies of Design Research*. Hong Kong: IASDR, 2007.

Redström, Johan.
"En Experimenterande Designforskning."
In *Under Ytan: En Antologi om Designforskning*, edited by Sara Ilstedt Hjelm, 164-177. Stockholm: Raster, 2007.

Redström, Johan.
"Persuasive Design: Fringes and Foundations."
In *Proceedings of the conference on Persuasive Technology for Human Well-Being (PERSUASIVE)*, 112-122. Berlin: Springer, 2006.

Masters theses

Danielsson, Sara.
"3 Steg för Nytt Stadsliv."
MA thesis in Design, School of Design and Crafts at Göteborg University, Sweden, 2005.

Ludvigsson, Mattias.
"Reflection through Interaction – Raising Energy Awareness Among Young People with Interaction Design and Speculative Re-design of Personal Objects."
MSc thesis in HCI/Interaction Design, IT University Göteborg, Sweden, 2005.

Müller, Carolin.
"Energy Curtain – A Self-Sustaining Curtain Using Photovoltaic Technology to Light Up Optical Fibres."
MSc thesis in Textile Engineering, Swedish School of Textiles at the University College of Borås, 2005.

Sylwan, Kerstin.
"Solparasit."
MA thesis in Design, School of Design and Crafts at Göteborg University, Sweden, 2005.

References about the design examples

The Element

Gyllenswärd, Magnus, Anton Gustafsson and Magnus Bång.
"Visualizing Energy Consumption of Radiators."
In *Proceedings of the conference on Persuasive Technology for Human Well-Being (PERSUASIVE)*, 167-170. Berlin: Springer, 2006.

Energy Curtain

Ernevi, Anders, Margot Jacobs, Ramia Mazé, Carolin Müller, Johan Redström and Linda Worbin.
"The Energy Curtain."
In *IT+Textiles*, edited by Johan Redström, Maria Redström and Ramia Mazé, 76-82. Helsinki: IT Press / Edita, 2005.

Müller, Carolin.
"Ambience for Energy Awareness: The Energy Curtain."
In *Proceedings of the conference on Intelligent Ambience and Well-Being (Ambience)*. Finland: Tampere University of Technology, 2005.

Müller, Carolin.
"Energy Curtain – A Self-Sustaining Curtain Using Photovoltaic Technology to Light Up Optical Fibres."
MSc thesis in Textile Engineering, Swedish School of Textiles at the University College of Borås, 2005.

Routarinne, Sara and Johan Redström.
"Domestication as Design Intervention."
In *Proceedings of the Nordic Design Research conference*. Stockholm: Konstfack / NORDES, 2007.

Erratic Appliances

Ernevi, Anders, Samuel Palm and Johan Redström.
"Erratic Appliances and Energy Awareness."
In *Knowledge, Technology & Policy,* Special Issue on Design Research 20, no. 1 (2007): 71-78.

Ernevi, Anders, Samuel Palm and Johan Redström.
"Erratic Appliances and Energy Awareness."
In *Proceedings of the Nordic Design Research conference*. Copenhagen: Danish Center for Design Research / NORDES, 2005.

Routarinne, Sara, and Johan Redström.
"Domestication as Design Intervention."
In *Proceedings of the Nordic Design Research conference*. Stockholm: Konstfack / NORDES, 2007.

Power-Aware Cord

Gustafsson, Anton and Magnus Gyllenswärd.
"The Power-Aware Cord: Energy Awareness through Ambient Information Display."
In *Proceedings of the conference on Human Factors in Computing Systems (CHI).* New York: ACM Press, 2005.

Ilstedt Hjelm, Sara, Magnus Gyllenswärd and Anton Gustafsson.
"Designing for Energy Awareness – The Power-Aware Cord."
In *Proceedings of the conference for Cultural Heritage and the Science of Design (CUMULUS)*. Lisbon, Portugal: IADE, 2004.

Löfström, Erica.
"Visualizing Energy in Households: The Power Aware Cord as a means to create energy awareness."
In *Proceedings of the European Network for Housing Research conference*. Rotterdam, The Netherlands: ENHR, 2007.

Löfström, Erica.
Visualisera energi i hushåll - avdomesticeringen av sociotekniska system samt individ respektive artefaktbunden energianvänding.
(Visualizing Energy in Households – The de-domestication of socio-technical systems and individually- as well as artefact-bound energy use.)
PhD dissertation, Department of Technology and Social Change, Linköping University, Sweden, 2008.

Löfström, Erica and Jenny Palm.
"Visualizing Energy Used in Households to Develop Sustainable Energy Systems – Different Methods."
In *Proceedings of the European Network for Housing Research conference*. Ljubljana, Slovenia: ENHR, 2006.

Free Energy

Jacobs, Margot and Ulrika Löfgren.
"Promoting Energy Awareness through Interventions in Public Space."
In *Proceedings of the Nordic Design Research conference*.
Copenhagen: Danish Center for Design Research / NORDES, 2005.

Jacobs, Margot, Ulrika Löfgren and Ramia Mazé.
"Free Energy: Alternative Designs for Awareness and Choice."
In *Proceedings of the conference for Cultural Heritage and the Science of Design (CUMULUS)*. Lisbon, Portugal: IADE, 2004.

References from the essay and perspectives

Allen, Stan.
"Its Exercise, Under Certain Conditions."
Columbia Documents of Architecture and Theory 3 (1993): 89-113.

Ambasz, Emilio, curator and ed.
Italy: The New Domestic Landscape, Achievements and Problems of Italian Design.
New York: Museum of Modern Art, 1972.

Bell, Jonathan.
"Ruins, Recycling, Smart Buildings, and the Endlessly Transformable Environment."
In *Strangely Familiar: Design and Everyday Life,* edited by Andrew Blauvelt, 72-88. Minneapolis, MN: Walker Art Center, 2003.

Berker, Thomas, Maren Hartmann, Yves Punie and Katie Ward.
"Introduction."
In *Domestication of Media and Technology*, edited by Thomas Berker, Maren Hartmann, Yves Punie and Katie Ward, 1-17. Maidenhead, UK: Open University Press, 2005.

Betsky, Aaron.
"The Strangeness of the Familiar in Design."
In *Strangely Familiar: Design and Everyday Life*, edited by Andrew Blauvelt, 14-37. Minneapolis, MN: Walker Art Center, 2003.

Blauvelt, Andrew, ed.
Strangely Familiar: Design and Everyday Life.
Minneapolis, MN: Walker Art Center, 2003.

Blythe, Mark, Kees Overbeeke, Andrew Monk and Peter Wright, eds.
Funology: From Usability to Enjoyment.
Dordrecht, The Netherlands: Kluwer, 2003.

Borgmann, Albert.
Technology and the Character of Contemporary Life.
Chicago, IL: University of Chicago Press, 1984.

Buchanan, Richard.
"Declaration by Design: Rhetoric, Argument, and Demonstration in Design Practice."
In *Design Discourse*, edited by Victor Margolin, 91-109. Chicago, IL: University of Chicago Press, 1989.

Chapman, Jonathan.
Emotionally Durable Design: Objects, Experiences and Empathy.
London: Earthscan Publishers, 2005.

Cross, Nigel, ed.
Developments in Design Methodology.
Hoboken, NJ: John Wiley & Sons, 1984.

Dobers, Peter and Lars Strannegård.
"Design, Lifestyles and Sustainability: Aesthetic Consumption in a World of Abundance."
Business Strategy and the Environment 14 (2005): 324–336.

Dunne, Anthony.
Hertzian Tales: Electronic Products, Aesthetic Experience and Critical Design.
London: Royal College of Art Computer Related Design Research Publications, 1998.

Dunne, Anthony and Fiona Raby.
Design Noir: The Secret Life of Electronic Objects.
Basel, Switzerland: Birkhäuser, 2001.

Fiedler, Jeannine, Peter Feierabend and Ute Ackermann, eds.
Bauhaus (Design).
Köln, Germany: Könemann, 2000.

Forlizzi, Jodi and Carl DiSalvo.
"Service Robots in the Domestic Environment: A Study of the Roomba Vacuum in the Home."
In *Proceedings of the conference on Human-Robot Interaction (HRI).*
New York: ACM Press, 2006.

Frayling, Christopher.
"Research in Art and Design."
Royal College of Art Research Papers 1, no. 1 (1993-4): 1-5.

Fry, Tony.
"Rematerialisation as a Prospective Project."
Design Philosophy Papers 3, 2004.

Gaver, Bill, Tony Dunne and Elena Pacenti.
"Cultural Probes."
Interactions 6, no. 1 (1999): 21-29.

Gaver, William, John Bowers, Andy Boucher, Andy Law, Sarah Pennington and Nicolas Villar.
"The History Tablecloth: Illuminating Domestic Activity."
In *Proceedings of the symposium on Designing Interactive Systems (DIS).*
New York: ACM Press, 2006.

Hunt, Jamer.
"Just Re:Do it: Tactical Formlessness and Everyday Consumption."
In *Strangely Familiar: Design and Everyday Life*, edited by Andrew Blauvelt, 56-71. Minneapolis, MN: Walker Art Center, 2003.

Laurel, Brenda, ed.
Design Research: Methods and Perspectives.
Cambridge, MA: MIT Press, 2003.

Margolin, Victor.
"The Product Milieu and Social Action."
In *Discovering Design: Explorations in Design Studies*, edited by Richard Buchanan and Victor Margolin, 121-145. Chicago, IL: University of Chicago Press, 1995.

Mazé, Ramia.
Occupying Time: Design, Technology and the Form of Interaction.
PhD dissertation in Interaction Design, Malmö University / Blekinge Institute of Technology. Stockholm: Axl Books, 2007.

Mazé, Ramia and Johan Redström.
"Difficult Forms: Critical Practices in Design and Research."
Research Design Journal 1, no. 1 (2009): 28-39.
Moggridge, Bill.
Designing Interactions.
Cambridge, MA: MIT Press, 2006.

Molander, Bengt.
Kunskap i Handling.
Göteborg, Sweden: Daidalos, 1996.

Norman, Donald.
The Invisible Computer.
Cambridge, MA: MIT Press, 1998.

Redström, Johan.
"On Technology as Material in Design."
Design Philosophy Papers, Collection Two, edited by Anne-Marie Willis. Ravensbourne, Australia: Team D/E/S Publications, 2005.

Redström, Johan.
"Aesthetic Concerns."
In *Pervasive Information Systems*, edited by Panos Kourouthanassis and George Giaglis. Armonk, NY: M.E. Sharpe, 2007.

Redström, Johan.
"Persuasive Design: Fringes and Foundations."
In *Proceedings of the conference on Persuasive Technology for Human Well-Being (PERSUASIVE)*, 112-122. Berlin: Springer, 2006.

Redström, Johan, Maria Redström and Ramia Mazé, eds.
IT+Textiles.
Helsinki: IT Press / Edita, 2005.

Robach, Cilla, curator and ed.
Konceptdesign.
Stockholm: Nationalmuseum, 2005.

Routarinne, Sara and Johan Redström.
"Domestication as Design Intervention."
In *Proceedings of the Nordic Design Research conference*.
Stockholm: Konstfack / NORDES, 2007.

Seago, Alex and Anthony Dunne.
"New Methodologies in Art and Design Research: The Object as Discourse."
Design Issues 15, no. 2 (1999): 11-17.

Shove, Elizabeth, Matthew Watson, Martin Hand and Jack Ingram.
The Design of Everyday Life (Cultures of Consumption).
Oxford, UK: Berg, 2007.

Silverstone, Roger.
Television and Everyday Life.
London: Routledge, 1994.

Silverstone, Roger, Eric Hirsch and David Morley.
"Information and Communication Technologies and the Moral Economy of the Household."
In *Consuming Technologies – Media and Information in Domestic Spaces*, edited by Roger Silverstone and Eric Hirsch, 15-31.
London: Routledge, 1992.

Stolterman, Erik.
"*Designtänkande.*"
In *Under Ytan: En Antologi om Designforskning*, edited by Sara Ilstedt Hjelm, 12-19. Stockholm: Raster, 2007.

Thackara, John.
"Beyond the Object in Design."
In *Design after Modernism,* edited by John Thackara, 11-34. New York: Thames and Hudson, 1988.

Verbeek, Peter-Paul.
"Devices of Engagement: On Borgmann's philosophy of information and technology."
Techné 6, no. 1 (2002): 69-92.

Willis, Anne-Marie.
"Ontological Designing – Laying the Ground."
Design Philosophy Papers, Collection Three, edited by Anne-Marie Willis.
Ravensbourne, Australia: Team D/E/S Publications, 2007.